RICE WITHOUT RAIN

Jinda, a teenage girl in Northern Thailand, has her quiet life suddenly disrupted when student activists from Bangkok come to her village. The students incite the villagers to rebel against the landlords who take away half of their crops every year. Jinda grows increasingly attracted to Ned, one of the students, and through their growing friendship she learns about politics, about love, and about herself as an independent young woman.

ABOUT THE AUTHOR

Minfong Ho was born in Rangoon, Burma, in 1951. She had a happy childhood living on the outskirts of Bangkok, Thailand, in a self-contained little compound. Minfong liked the ''usual things'' — chasing catfish in the rain, roasted coconuts and fried bananas. Minfong was educated at the International School of Bangkok; Tunghuai University, Taichung, Taiwan, and Cornell University, Ithaca, New York.

From 1975–77 she was a lecturer at Chiengmai University, Chiengmai, Thailand; from 1978–80 she was a teaching assistant in the English Department at Cornell University, New York, and in 1985, she was Writer-in-Residence at the University of Singapore.

She has also been a journalist, a legal assistant, a social worker, and a nutritionist and a relief worker on the Thai–Cambodian border. She was married in 1976 and has two children, Donnah Donfung Dennis and Mary Schuofung Dennis.

Minfong says she wanted to write *Rice Without Rain* because she ''wanted to understand what happened to the Thai student movement during 1973–76, why it succeeded, why it failed, how it affected Thai farmers''.

RICE WITHOUT RAIN

MINFONG HO

*for those who died at Thammasart on October
6, 1976, and those who survived to help build a
better Thailand, this book is respectfully
dedicated.*

HEINEMANN
NEW WINDMILLS

Heinemann Educational Books Ltd
Halley Court, Jordan Hill, Oxford OX2 8EJ
OXFORD LONDON EDINBURGH
MADRID ATHENS BOLOGNA PARIS
MELBOURNE SYDNEY AUCKLAND
IBADAN NAIROBI HARARE GABORONE
SINGAPORE TOKYO PORTSMOUTH NH (USA)

ISBN 0 435 12340 8

First published in 1986 by André Deutsch Ltd
First published in the New Windmill Series 1989

91 92 93 94 95 10 9 8 7 6 5 4 3

Cover illustration by Alison Clare Darke

Printed in England by Clays Ltd, St Ives plc

Like the rice, we live in wait for the rain,
In times of drought, we wither in the fields.
How many of us must die
To feed those who're already fat?
Like the rice, like the withered rice,
We live in wait for the rain.

(adapted from a Thai folksong)

Chapter 1

Heat the colour of fire, sky as heavy as mud, and under both the soil — hard, dry, unyielding.

It was a silent harvest. Across the valley, yellow rice fields stretched, stooped and dry. The sun glazed the afternoon with a heat so fierce that the distant mountains shimmered in it. The dust in the sky, the cracked earth, the shrivelled leaves fluttering on brittle branches — everything was scorched.

Fanning out in a jagged line across the fields were the harvesters, their sickles flashing in the sun. Nobody spoke. Nobody laughed. Nobody sang. The only noise was wave after wave of sullen hisses as the rice stalks were slashed and flung to the ground.

A single lark flew by, casting a swift shadow on the stubbled fields. From under the brim of her hat, Jinda saw it wing its way west. It flew to a tamarind tree at the foot of the mountain, circled it three times and flew away.

A good sign, Jinda thought. Maybe the harvest won't be so poor after all. She straightened up, feeling prickles of pain shoot up her spine, and gazed at the brown fields before her. In all her seventeen years, Jinda had never seen a crop as bad as this one. The heads of grain were so light the rice stalks were hardly bent under their weight. Jinda peeled the husk of one grain open: the rice grain inside was no thicker than a fingernail.

Sighing, she bent back to work. A trickle of sweat ran down between her breasts and into the well of her navel. Her shirt stuck to her in clammy patches, and the sickle handle was damp in her palm. She reached for a sheaf of rice stalks and slashed through it.

Reach and slash, reach and slash, it was a rhythm she must have been born knowing, she thought, so deeply ingrained was it in her.

Out of the corner of her eye, she saw the hem of her sister Dao's sarong, faded grey where once the bright flowered pattern had been. Dao was stooped even lower than the other harvesters in their row, and was panting slightly as she strained to keep up.

From the edge of the field came the sudden sound of a thin, shrill wail.

'Your baby's crying, Dao,' Jinda said.

Her sister ignored her.

'Oi's crying,' Jinda repeated. 'Can't you hear him?'

'I hear him.'

'Maybe he's hungry.'

'He's always hungry.'

'Why don't you feed him, then?'

'Why don't you mind your own business?' Dao snapped.

'But couldn't you try?' Jinda insisted, as the wailing got louder. 'I think you should at least try.'

Dao slashed through a sheaf of stalks and flung them to the ground. 'When I want your advice, sister,' she said, 'I'll ask for it.'

They did not speak again for the rest of the afternoon. The baby cried intermittently, but Jinda did her best to ignore it.

How different this is from past harvests, Jinda thought. Just three years ago, before the drought, she and Dao had gaily chatted away as they cut stalks heavy with grain. They had talked about what they might buy after the harvest — new sarongs, some ducklings, a bottle of honey. And as they talked, the dark handsome Ghan had sung love songs across the fields to Dao, until her face turned so red she had to run down to the river and splash cold water on it.

Ghan and Dao were considered a perfect match by the whole village. After all, weren't their fathers two of the most important men in Maekung? True, there was a deep, silent hostility between Dao's father, the village headman, and Ghan's father, the village healer, but that only seemed all the more reason why a union between their children would be auspicious.

So when Dao and Ghan were married, the whole village attended the wedding. All that morning the hundred or so families of Maekung each took their turn to tie the sacred thread around the bridal couple's wrists, and after the elaborate wedding feast, countless couples, young and old, had danced the Ramwong until the moon rose high above the palm trees and the kerosene lanterns were lit.

There had been so much of everything then, Jinda thought wistfully, so much food and rice wine, so much music and movement, and best of all, so much laughter.

And now, just two very poor harvests later, there was never any laughter, nothing but the whisper of sickles against dry stalks in parched fields. Ghan had left to work in the city even before their son was born, and Dao — poor Dao, Jinda thought, stealing a glance at her sister's grim face — Dao had become just a shadow of herself.

When twilight finally came, the line of harvesters broke up, and the men and women straggled back to the edge of the fields. Most rested under the shade of little thatched lean-tos, fanning themselves with their hats, while others ladled water from rusty buckets and drank deeply.

Jinda tucked her sickle into the waist of her sarong and

walked up to her sister. 'Want to go down to the river?' she asked.

'Too tired,' Dao said, massaging her back with one hand.

'You can stretch out under the banyan tree by the river bank.'

'And the baby?'

'Bring him, of course. He likes the cool water, and you like bathing him. And me, I like watching the two of you together.'

Dao smiled then, and Jinda knew their fight was over. 'All right,' Dao said, 'I'll go and get the baby.'

As Jinda watched her sister stoop to enter the temporary lean-tos where other harvesters were working, her father walked up to her.

'Tired?' he asked quietly.

Jinda smiled, shaking her head. 'You're the one that should be tired, Father,' she said. 'I saw you helping Lung Teep carry his loads of cut rice stalks when you should've been resting.'

'Teep's leg is still sore,' Jinda's father said, rubbing his own shoulder ruefully.

Jinda longed to massage her father's shoulders for him, but knew that she was too old to do that now. There had been a time years ago, shortly after her mother had died, when she used to massage his shoulders every evening after he came back from the fields. He'd sit on the top step of the porch, and she'd kneel behind him, deftly easing away the tight knots from his shoulders. He'd talked quietly of the day's work, much as he had with Jinda's mother before she died. After Dao had married and moved away to live in her father-in-law's house, Jinda had grown even closer to him, taking over the chores of preparing the meals and laying out the sleeping mats that Dao used to do.

'Could we have dinner a little late tonight, Father?' Jinda asked now. 'Dao and I are going down to bathe in the river.'

Inthorn frowned. 'Why don't you two forget about the river today,' he said.

'Why? Are you very hungry?'

'I'd rather you come straight home, that's all,' the farmer said.

'But why?'

Inthorn hesitated. 'It's getting dark,' he said.

Jinda laughed in surprise. 'But it's no darker than when we usually go.'

'Still, you never know what strangers you might meet there.'

'Strangers? Father, really! I've never met a stranger at the river in my whole life, and you know it! What are you talking about?'

For answer Inthorn pointed west, where the setting sun glowed behind the Chiangdao mountain range. Silhouettes of gnarled teak trees stood against the cloudless sky, stark and motionless.

'I don't see anything,' Jinda said.

'On the nearest ridge,' her father answered, pointing. 'Halfway down the Outbound Path, can't you see?'

The Outbound Path linked their small valley to the road through the mountain range, and to the big highway leading to the bustling town of Chiengmai nearly fifty miles away. Curious, Jinda scanned the winding path.

'There's nobody . . .' Jinda began, then broke off. She saw a group of three or four people, carrying what looked like heavy knapsacks climbing down the hillside.

'How strange,' Jinda said softly. These people were walking on foot. The few pedlars who came always rode motorcycles, and the rent-collector always roared in on his truck. Even the poorest farmer would have used a bullock cart or bicycle to travel that long path into their valley. 'They're walking down,' Jinda said. 'Who could they be?'

Inthorn shook his head. 'I don't know,' he said, 'but I don't like it.' Jinda thought of the recent radio reports they had heard, of Communist insurgents in the area, and understood the tension in her father's voice. He patted her shoulder awkwardly and said, 'So, how about going straight home and starting dinner, eh?'

Before Jinda could agree, Dao stepped out of the lean-to, holding the baby in her arms. 'Little Oi's ready for a cool bath, Jinda,' she laughed. 'Let's go! I'm really looking forward to it.'

Jinda and her father exchanged a quick look. It was not often that Dao looked forward to anything these days. 'All right, go ahead,' Inthorn said reluctantly. 'But be quick.'

As Dao walked on ahead, Jinda turned back to look down the Outbound Path. Even in the dimming light, the strangers kept a brisk pace, and were approaching the foot of the mountain. At this rate, Jinda thought nervously, they would reach the river just before nightfall. She quickened her own step, and hurried past her sister.

Feathery ashes from brush fires on the mountainside had blown down, speckling the bamboo leaves fallen on the path below. At the end of this path was the river, a ribbon of cool silver under the bamboo groves.

Jinda reached the river first and stood on the bank waiting for Dao. It was shady there, screened off from the afternoon sun by a web of branches. A salamander slithered across the path and into a patch of ferns.

When Dao arrived, she laid her baby down on the river bank. Turning away from Jinda, she unbuttoned her shirt and peeled the sweaty cloth from her back. With a deft twist, she untied the knot of her sarong and pulled it up over her breasts. Wrapping the cloth tightly around her, she knotted the sarong.

Jinda shrugged off her own shirt, and knotted her sarong around her, aware that her own breasts were still smooth and firm when her sister's had already started to sag.

She watched Dao cradle the baby as she waded into the river, tiny ripples fanning from her ankles. In midstream, where it was now only knee-deep because of the drought, Dao carefully sat down, hugging the baby to her. The clear water tinted her sarong a deep green.

'It's cold!' Dao laughed, and for an instant she looked lovely and carefree again.

Jinda waded out and sat down next to her. The baby was lying half immersed in the cloth hollow of her sister's lap, as Dao trickled handfuls of water over him. The baby cooed, shining sleek and slippery in the sunlight, and Dao crooned back to him. She pulled him gently to his feet, where he wobbled on thin legs for a second, before collapsing with a splash back into Dao's lap.

'Poor Oi, poor little Oi,' she said softly. 'Why can't you stand straight?' She ran her hands over his legs, so frail they looked like twigs. 'Look at you, so skinny now,' she murmured, 'when newborn, you were soft as dough. And your hair — why has it turned so dry and brown, like straw?' She stroked the hair on his big head, careful not to tug at it because tufts had come off in her hands recently.

Jinda watched in silence as Dao slid her hands down the baby's neck and over the stomach that bulged below his ribs. 'And your belly,' she said. 'It's the only part of you that keeps growing while the rest of you keeps shrinking. Why, Oi, why?' And although her voice was still gentle, the gaiety was gone from it.

'It's not because you're hungry, little one, is it? I feed you spoon after spoonful of rice gruel, until you don't want a drop more, so how could you be hungry? Is it milk you want, then? Is it?'

She loosened her sarong, and turned slightly away from Jinda. Cradling the baby in her arms, she guided him towards her right breast.

Jinda held her breath. She tried not to look, knowing that her sister had become shy about breast-feeding ever since her milk had begun to dry up. For the first four months Dao had suckled her baby openly, but as the baby's appetite increased, Dao grew thinner and her milk supply lessened. No matter how frequently she'd try to breastfeed little Oi, he would always fret for more, kneading her breasts with tiny fists.

He was whimpering now, his face screwed into a pout as he sucked at his mother's breast. But it was no use. There was no milk. He started to howl.

With a deep sigh, Dao plucked the baby away from her nipple and rocked him soothingly against her. Gradually he calmed down, his cries subsiding into little hiccups.

'I'm sorry, baby,' Dao told him, 'Please understand. I can't help it. I just don't have enough milk.' The baby gazed back at her with hollow, long-lashed eyes.

Just then a single sunbeam filtered down from the mesh of branches above them, and shone on Dao's face. The baby gazed at it in fascination. He reached out a shaky arm to touch the patch of sunlight, patting his mother's dappled cheek.

Dao laughed with delight. 'Jinda, look!' she called. 'He understands! He's patting me, trying to comfort me.'

'But he's only . . .' Jinda began in protest, then stopped.

There had been a sudden movement behind the branches of the banyan tree on the far side of the river, and Jinda thought she had also heard something.

'What is it?' Dao asked in alarm.

A stranger stepped out of the shadows and stood, towering over them, on an overarching root of the banyan tree. Slowly he raised both hands, palms together, and bent his head towards them, in the traditional Thai greeting.

'Sawadi,' he said, and although his voice was deep and solemn, there was a faint trace of a smile on his lips.

Jinda returned neither the greeting nor his smile. 'What do you want? Who are you?' she asked sharply.

'I'm sorry if I startled you. We mean no harm.' The stranger glanced behind him, and Jinda saw three figures in the shadows behind him. 'We just wanted to know if this is the village of Maekung?'

'What if it is?' Jinda said.

'We'd like to spend tonight — perhaps longer — here, that's all.'

He spoke slowly, but with a distinct urban accent which Jinda found disquieting.

The baby began to whimper, and Dao, shivering now, hugged him more tightly to her.

'I think you'd better leave,' Jinda said.

'Why?' the stranger smiled, and his teeth gleamed in the twilight.

The baby was crying loudly now, and Dao looked scared.

'You're not welcome here,' Jinda said.

'Isn't that a decision only the headman of the village should make?' The stranger asked, still smiling. He jumped lightly from the tree-root into the water, and waded towards them. His shirt was unbuttoned, and his bare chest glistened with water in the twilight. 'Who are you,' he asked Jinda mockingly, 'to speak for the village headman?'

Jinda stood up and drew herself to her full height. 'I am Jinda Boonrueng,' she said evenly. 'His daughter.'

For a second he stared at her. Then he averted his eyes, but not quickly enough to hide the sudden flash of interest in them.

Suddenly self-conscious, Jinda pulled her wet sarong higher over her breasts and turned away abruptly. With as much dignity as she could manage, she helped Dao up, and together the two sisters waded back to their side of the river bank. There, they gathered their strewn clothes, and hurried down the path towards the village.

Jinda never once looked back, but she thought she could feel his gaze burn into her. No one had ever looked at her like that before, she thought, with such intense interest. She felt strangely exhilarated, and could barely keep from laughing out loud.

It was only when she was almost home that it occurred to Jinda that this stranger's interest in her might well have been sparked only because she was the village headman's daughter, and not because she was strikingly beautiful in a wet sarong.

'What a stupid fool!' Jinda blurted out.

'Who?' Dao asked behind her.

Jinda stopped in confusion. 'Nothing . . . no, nobody,' she stuttered.

Her sister looked at her curiously. 'You couldn't have meant that young man at the river,' Dao said. 'The way he talked, he was no fool. He had a fancy Bangkok accent. Didn't you notice?'

Jinda shook her head. 'No I didn't,' she said, adding, 'and I don't care.' Then she flushed, realising how obvious it must be to her sister that both statements were total lies.

Chapter 2

Jinda ran a quick hand down her thigh. The sarong was still slightly damp, and felt pleasantly cool against her skin. She had put on a fresh white blouse, and was now twisting her long straight hair into a neat bun.

It's not as if tonight was the Loy Krathong festival, Jinda thought, why am I bothering with my hair? Her reflection smiled at her from the little mirror on the veranda. Because you want to look pretty, it told her.

Jinda had never been very interested in looking pretty. She knew, studying the mirror now, that her eyes were big, her eyelashes long, and her complexion smooth and fair. A typical Northern Thai beauty, Jinda told herself, and grinned. It was her grin that spoiled it. She had been told often enough to smile demurely ('Don't laugh with all your teeth showing!' Dao would say), but she couldn't help it. Other village girls learned to smile because it made them pretty; Jinda laughed when she thought something was funny, which was far too often for her sister's taste.

Her father had never minded, though. Hadn't he always treated her like the son he had yearned for before little Pinit was born? He had taught her to fly kites, to wield an axe, and to read all the old newspapers he read, so that even though, like the other village girls, she had only attended the district school for four years, she could read and write better than many of the men in the village. And her father took great pride in that.

Yet here I am, Jinda thought, preening before the mirror just like giggly cousin Mali. I'd probably tuck a white gardenia in my hair if there were any in bloom. Jinda grinned again, and deftly pinned her hair back at the nape of her slender neck.

Pausing to pick up the little oil-lamp at the edge of the verandah, Jinda hurried down the stairs and towards the temple. She was late, and the temple was on the outskirts of the village, a good ten minutes' walk away. Her father, grandmother and little brother had all headed there as soon as they'd heard some neighbours announce the presence of four strangers in the temple courtyard. Jinda, using her wet sarong as an excuse, had lagged behind to change and comb her hair.

A cool evening breeze blew against her bare arms, and Jinda shivered, then shielded the single flame of her lamp with her hand. It was early November, and the night air was getting chilly.

In the distance she could hear the sweet tinkling of the tiny bells hung high on the wooden temple eaves. A crescent moon was suspended above the mountains, as a few fireflies darted among the orange blossoms of the flame-of-the-forest trees.

As Jinda approached the temple, she saw that a fire had been lit in the courtyard. The light of its flame flickered on the brick walls, and cast into relief the sweeping temple roof arching above. Quite a few of the villagers must have already gathered here, Jinda thought, for there to be such a big bonfire.

She slipped past the temple gates, and into the courtyard. Around the fire were about fifty people. Wrinkled old faces inset with eyes that glittered like bits of orange tile from the temple roof, glossy-faced boys with their shoulders wrapped in homespun shawls, young women hugging babies limp with sleep against them — everyone waited silently by the fire.

Set apart from them, on the other side of the fire, Jinda saw the four strangers. One of them lifted his head as she edged her way towards the fire, and nodded towards her.

Jinda caught her breath. He looked even more striking than he had at the river. Calm and at ease despite the stares of the crowd, he sat there, legs tucked neatly under him like a cat, watching her.

'Now that we're all here,' he said across the fire, 'I would like to begin the meeting.'

For a second Jinda thought he was talking to her, but then realised that he was really addressing her father, who was seated near by. Relieved, she quietly settled down at the edge of the crowd.

'Please start,' Inthorn said formally. 'We would be pleased to listen to what you have to say.'

'I had thought that I would start by introducing each of us to you,' the stranger said, and his gaze seemed to seek Jinda out and focus on her even though she was at the very edge of the crowd. 'But now I think that perhaps the best way to start, unusual as it may be, is to sing to you a song which would tell you about us better than any words can.'

He smiled, and to Jinda it seemed as if he was smiling directly at her. Then, taking a deep breath, he began to sing.

His voice was slow and deliberate, almost a chant, and in the stillness it spiralled up to the trees canopying the fire. One by one, hesitantly at first and then with more vigour, the three other strangers joined him in song, until it seemed as if the song was surging forth from the night itself.

Jinda hugged her knees to her and listened, entranced. She had never heard anything like this in her life, yet the song seemed hauntingly familiar.

The song was the song of rice. It sang of the sowing of the unhusked seed rice, and of the careful transplanting of the seedling from seedbed to the newly ploughed fields. It sang of the months of weeding, of watering, of waiting as the stalks grew tall and green, then ripened into a dry brown. It sang of the days spent harvesting and threshing, milling and winnowing the grain, for a single ricebowl to be filled.

The melody and the words rippled outwards, filling the cool night air, until the last note died away, sharp and sweet.

Jinda sat absolutely still. No, she had heard nothing like this before, and yet she felt as if she had known the song all her life. How could this be?

Nor was she the only one to wonder. She saw her father reach over and take the stranger's hand, and hold it in the firelight. The farmer turned the student's palm towards the fire, and looked at it for a long moment.

'I don't understand,' Inthorn said at last. 'Your hands are smooth and soft, a student's hand. Yet your song,' he shook his head slowly, 'it sings of our life. How could you know what it's like for us? Where did you learn the song? Who are you?'

'We're students from Bangkok, just like the man who wrote that song was. His name was Jit Pumisak, and he chose to live among the landless farmers in the Northeast. He worked with them and for them, and his writings have inspired thousands of us to want to do the same.' He scanned the circle of faces around him, and smiled. 'I'd like to introduce each of us here to you now,' the stranger said, and although he spoke quietly, his voice carried well, and Jinda could hear him clearly. 'Jongrak and Pat,' he gestured to the two young men on his left, who nodded, 'are from Chulalongkorn University. And Sri beside me here, is a medical student from Mahidol.' The person on his right looked up and smiled shyly. Jinda saw with surprise that, despite the short hair and jeans, it was a girl. A rather pretty girl, too, Jinda noted with a twinge.

'As for myself,' the stranger continued, 'I'm a third-year student at Thammasart University.'

'And your name?' someone from the crowd prompted.

'Ned,' he said.

'Just . . . Ned?' the question sounded sly, as if it was a trap. Jinda turned and saw Mau Chom, the village healer. Sitting beside him, frowning slightly, was her sister Dao, his daughter-in-law. Jinda wished Dao had sat next to their own father instead. 'My full name is Nedmanoon Angkul-prasert,' the stranger answered easily, 'but I doubt that anyone would have any occasion to use it.'

A girl giggled appreciatively, and Jinda saw Mali, sitting very close to the fire. And with a flower in her hair no less, Jinda thought wryly.

Ned continued to explain that all four students were from urban backgrounds, and knew very little about village life. 'We feel that as students,' he said, 'and as loyal citizens of Thailand, our education wouldn't be complete unless we learned more of how the farmers of our country lived. We have about two months before the new term starts, and during this time we'd like to stay here in Maekung with you.'

There was a pause, as Ned waited for Inthorn's response. Jinda noticed that her father was cracking his knuckles, something he did when he was especially worried or puzzled. 'It is unusual, what you have asked,' he said. 'No one, and certainly not university students from Bangkok, has ever just wanted to visit and stay here before. What would you do here?'

'We'd work alongside you, sir. We'd try to learn what sort of problems you might have. Perhaps we could even try to help in small ways. Sri here has brought medical supplies, for example . . .'

'That's all very well,' Inthorn said, 'but why would you want to waste your time doing this?'

'As I said, sir. We want to learn more about the countryside of Thailand.'

'But why? What good will it do you? Who are you?'

'But I just told you, sir!' Ned said. 'What more do you want to know?'

'He wants to know,' the thick hoarse voice of Mau Chom broke in, 'Whether you are Communists.'

There, it was out. The word which had lurked in every villager's mind as soon as the rumour spread that strangers were coming to Maekung. 'Communist' — the word broadcast over the army radio stations to warn of bloodshed and depravity.

And just last month, hadn't the radio been full of news of strange battles in Bangkok, when masses of students had rebelled against the government, when the tanks had been called out and hundreds gunned down and killed? Of course no one in Maekung had paid much attention to this. It was only Politics, games periodically played by politicians who never left the city. But Communists were more real than politicians: Communists actually lived in rural areas, and ventured into villages. Didn't they raid village granaries, and behead children and burn whole villages to the ground?

Jinda looked at the four strangers sitting by the fire. They all looked dusty, dishevelled, and very tired. The small pale girl called Sri was actually nodding off to sleep, her thin shoulders slouched. They certainly didn't look as if they'd be beheading many people tonight, Jinda thought.

'No, we are not Communists,' Ned said, looking Mau Chom straight in the eye. 'Nor do we have sub-machineguns packed in our knapsacks. Any more questions?'

Even in the half-light of the fire, Jinda could see the spirit doctor's face darken. He muttered something under his breath which Jinda couldn't hear, but which she was sure was hostile. One of the other students quickly spoke up. 'Please excuse brother Ned for speaking impatiently,' he said. 'But we have been travelling for two days, by overnight train from Bangkok to Chiengmai, and then by bus and by foot all today. We're all very tired.'

Ned nodded, and said to Inthorn apologetically, 'Perhaps it was too much, sir, and too soon, to ask if we could spend eight weeks in your village. But could we at least sleep here tonight?'

Inthorn glanced at the glowering Mau Chom, and then turned back to Ned. 'The two monks at the temple don't have enough room in their quarters to accommodate you,' Inthorn said, 'but we'd be happy to share our food and shelter with you tonight.'

Ned's smile was one of sheer relief, and for the first time, Inthorn smiled back at him. 'Welcome to Maekung,' the head of the village said.

An excited murmur rose from the crowd. It was as if the village had been holding back its own welcome, until Inthorn had voiced it.

'Our homes are small,' Inthorn was saying now, 'and it will be easier to house you separately.' What he did not say, and what Jinda knew he meant, was that everybody's food supply was limited, and no one family could be expected to feed all four visitors at once.

Inthorn selected three farmers from the crowd and asked each one if they would house a student for the night. Each agreed. Finally he turned to the old woman sitting next to him and said, 'And you, Mother? Is it all right to welcome one of these young strangers to our house?'

Jinda's grandmother plucked the homemade cigar from her mouth and tucked it behind her ear. 'Of course it is,' she said, and grinned. 'Let's take that poor skinny one who's about to fall asleep right into the fire.'

Ned leaned over and gently shook Sri's shoulder. With a start, the student jerked awake. She took off her glasses, blinked, then stared bewildered, around her. Her cheeks were flushed from sleep and from the fire, and she looked delicate and vulnerable. 'What . . . where . . . oh dear, I'm sorry,' she stammered.

'Good Lord, he's a girl!' Jinda's grandmother exclaimed.

The crowd laughed, and Jinda shook her head. How could her grandmother have mistaken that very feminine-looking Sri for a boy, Jinda wondered? The old lady's eyesight must be worse than she'd admit.

'I'm sorry,' the girl said again. She was blushing furiously, the blood vessels fanning out on her smooth cheeks like tiny red ferns. 'I must have dozed off for just a moment. You were talking about, let's see, Communists, right?' she asked Ned.

Ned shook his head ruefully. 'That's precisely what I didn't want to talk about,' he said.

'Why not?' asked Jinda's grandmother quickly. Her eyes may be getting dim, Jinda thought, but her mind is as sharp as ever.

Ned hestitated. 'It's just that the 'Communist' question is a particularly sensitive one,' he said slowly. 'When we were organising the mass demonstrations in Bangkok, the military dictatorship labelled us Communists, and shot hundreds of us. Then later when many student groups, like this one, returned from visits to the countryside, they said some farmers would label them Communists, and shun them. It's just a crude kind of name-calling. No, we're not Communists,' Ned said.

'Then why do people keep calling you that?' the old woman persisted. She puffed at her cigar, and watched the smoke spiral upwards. 'Only a lit cigar gives off smoke,' she observed meditatively.

'Well, we have read some of their writing,' Ned admitted. 'And we even share a few of their ideas.'

'What ideas?' Lung Teep asked. There was a wariness in his voice, but a genuine curiosity too.

'That farmers who till the land should own it, for example,' the student said.

'That doesn't seem like such a bad idea to me,' Lung Teep said. 'You sure it's Communist?' He laughed, and a few other villagers laughed softly along with him.

'What other ideas do you have?' another farmer asked now.

'That the land rent is oppressively high,' Jongrak, one of the other students, spoke up eagerly. 'The rent should be lowered and yields more equitably distributed.'

'And there should be a democratically elected government representative of the people,' the other student chimed in.

'And a public health care system,' Sri added breathlessly.

Someone laughed, a full, throaty laugh which Jinda immediately recognised as her grandmother's. Her face tanned and wrinkled like an orange peel left out in the sun to dry, the old woman peered across the fire at the students. 'You all sound like old Buddhist monks chanting their Pali scriptures,' she said. Jinda could hear sympathetic chuckles from the crowd. They had not understood a word either.

Ned straightened up then, and stretched. That one slight movement drew everybody's attention back to him. 'We have many ideas for Thailand,' he said quietly. 'Some of them have long names, like social justice and land reform. But the basic idea behind them all is very simple.' Ned leaned back and looked at Jinda's grandmother, as if he was speaking only to her. 'It's as simple as ploughing the fields after a bad harvest, or digging a new well in search of fresh water when the old well has dried up. What we want for our country, grandmother, is the chance to start over again.'

It was quiet. Across the dying flames the old woman gazed at Ned. Age had pared away superfluous layers of flesh from her face and left a stark simplicity of skin over bone. As Jinda watched, her grandmother broke into a slow smile, the crinkles fanning out at the corners of her eyes.

'I like you,' she told Ned. 'You're so young.'

Chapter 3

The next morning, when the sky was barely light, Jinda got up. She wanted to start the cooking fire and have a good breakfast ready for their new guest. Jinda gave the pot a quick stir, and wished that she had some fish, or better yet, a little pork, to put in the stew. Still, with the lemon grass and basil, even the strips of eggplant smelled nice.

Across the open verandah, in the main part of the house, the others were stirring. Jinda could hear her grandmother putting away the sleeping mats and mosquito nets, and caught a glimpse of the old woman climbing down the verandah stairs. Sri, Jinda saw with surprise, was right behind the old woman, looking slightly rumpled, but as pale and delicate as the evening before.

'Come, I'll show you around,' Jinda heard her grandmother say with a touch of pride, 'You say you've never been in a village before?'

'Never,' Sri answered.

'Poor child, always stuck inside the schoolroom, weren't you? Never had the chance to go anywhere at all?'

'Well, actually,' Sri said hesitantly, 'I spent last summer in Europe, and the holiday before that in New York.'

Jinda shook her head. She wasn't sure at all she was going to like this girl.

Then Jinda heard her father go downstairs, to feed the buffalo tethered below the house. Last of all to get up was little Pinit, his light footsteps pattering across the wooden floorboards of the verandah. Jinda called out to him.

'Draw some more water from the well, Pinit!' she said. 'The urn up here is almost dry. And fill the drinking jars too, will you?'

Pinit stuck his head in the kitchen. 'Can't I do that later, sister?' he said. 'I want to watch that new girl with Granny.'

Jinda always found it hard to refuse her five-year old brother anything. 'Well, all right,' she said, 'but don't get in their way.' She wanted to add, 'And be sure to come back and tell me all about her!' I'm getting to be a regular old gossip like Aunty next door, she thought, and laughed.

Jinda was soaking the rice and wishing that the rice grain wasn't so small and broken, when Pinit burst through the door.

'Sister, come quick!' he panted. 'It's Granny! Hurry!'

Jinda dropped her ladle and dashed out of the kitchen door after him. 'What is it?' he asked. 'Is she hurt? Did she fall again?'

For answer Pinit pointed dramatically across the verandah towards the well.

Their grandmother was under the mango tree there, standing stock still, with one hand stretched out in front of her.

'I see a butterfly!' she shouted, 'It's got yellow stripes. Oh, my goodness, isn't it beautiful?'

Jinda rushed down the stairs and ran towards the old woman. It was only when she was a few feet away that she noticed her grandmother was holding something in front of her eyes. It glinted in the morning sunlight.

'Granny, what is it? What're you doing?' Jinda asked.

Then she saw the student, squinting near-sightedly at her, and realised what had happened.

'I can see!' The old woman said. 'These eye-rings of this child here, I've always wanted to try one on. They're amazing, Jinda, really they are. They sharpen the shapes of everything.' She turned towards Jinda and adjusted the slant of the glasses this way and that, peering through them all the while. 'It's best this way,' she said, clamping them on the tip of her nose and peering through them with her whole face tilted up. 'There now. Sharp, very sharp.'

Then she hobbled back to the hibiscus hedge, and picked a blossom from it. Carefully, she peeled each petal off, pulled out the long yellow pistil from it, and stuck that on her nose. It was a game she had played with Jinda years ago, to see who could balance a sticky hibiscus pistil on her nose the longest. Face upturned, the old woman laughed in delight, balancing both the drooping pistil and the glasses on the tip of her nose.

'Oh Granny!' Jinda laughed, and grasped the old woman's hands in her own. Hand in hand, with Sri groping her way close behind, they walked out of the gate and down the street.

In front of Sakorn's house, Jinda's grandmother paused. Adjusting the angle of the glasses, she studied the sagging roof and broken fence. 'Has that Sakorn grown so lazy,' she said, 'that he doesn't even repair his house any more?'

A few steps later, she stopped again. 'And why,' she asked, 'Is Sagnad's storage barn crumbling? Isn't he worried the rats will get at the rice? And what's happened to his wife's spice garden? Why is it all dried up?'

Jinda tried to pull her away, but the old lady stood firm. She turned around and confronted Sri. 'What magic is in these eye-rings of yours?' she asked Sri. 'If it's bad magic, why do I see the hibiscus, and the butterflies and clouds and mountains?' She shook her head, bewildered by the logic of her thoughts. 'But if it is good magic, why do I see my friends' homes so run-down and neglected?'

'It's not magic, grandmother,' Sri said gently.

'What is it, then?'

Sri hesitated. 'Optics, I guess,' she said.

The old woman grunted. 'Some people call crows ravens,' she retorted.

Leaving the two girls behind, she walked back through the swinging bamboo gate into her own yard. In a corner of it stood a little spirit house on its pedestal. The miniature wooden house had once been sturdy, with a gilt-edged roof and varnished walls. But the roof had caved in now, and its walls were bleached and warped. A jar of joss-stick stubs stood next to some withered jasmine buds on the platform in front of it. These were the only offerings.

Leaning on her cane, the old woman peered up at it. 'No rice, no candles, no fresh fruit?' she asked, and her voice trembled. 'But Jinda, what would your mother say? You promised her you'd care for the spirit house after she died. Why haven't you, child? Why aren't there even any candles?'

Jinda swallowed hard. 'We can't afford to buy any, Granny,' she said.

'But some bananas, a few tangerines? A lime or two?'

'The neighbourhood children would steal them from the altar,' Jinda said. 'So I stopped putting any there. There's a drought, Granny, you know that.'

'Surely a handful of rice . . .?'

'Father said that we don't have enough for ourselves,' Jinda said shakily. 'There's none to spare for the spirits.'

There was a moment of strained silence. Then, wordlessly, the old woman took off the glasses and handed them back to Sri.

In silence they walked back home. With her glasses back on, Sri looked around her with interest. She walked up to their house, and examined it closely. Jinda became acutely aware that their house, like the spirit house, was in disrepair. The thatched roof sagged, and its walls were bleached a bone-grey by the sun. The thick pilings on which the house rested were worm-ridden now, scarred by criss-crossing ruts where termites had chewed tunnels and laid their eggs. The wattle shed where the rice was stored had gaping holes gnawed by rats, which Inthorn hadn't bothered to repair because there was no longer any rice inside. Even the pig-pen under the rice-barn, once so full of squealing, squirming piglets, was now deserted.

'So this is your home,' Sri said quietly.

Home? The word sounded strained to Jinda. She realised that, for the first time in her life, she felt ashamed of her home.

The smell of the lemon grass stew simmering in the kitchen reminded Jinda that breakfast was almost ready, and she felt slightly better. 'Come up for some food,' she said, and started climbing up the stairs to the verandah.

The floorboards there were spotlessly clean, and the water jar almost full. So what if the chicken coop and the pigpen were deserted? It wasn't our fault that we were forced to sell all the animals because of the drought. We've done our best, Jinda told herself, we have nothing to be ashamed of.

And yet, she could not help but resent the careful way Sri walked across the porch, as if afraid the planks might give way under her. When Jinda offered her a drink of water from the earthenware jar, Sri asked if the water had been boiled first, then politely refused it when she heard it came straight from their well.

'I'll help you boil some water later,' she said.

Go and boil yourself, Jinda felt like saying. Instead, she stalked into the kitchen, and furiously stoked the cooking fire.

Sri followed her into the kitchen, and stood for a while by the open window. Outside, a few ducks were waddling to the river, their sleek feathers brushed glossy shades of brown by the morning light. The clean rich smell of the grass fire under the hayloft wafted up. Next to it, the buffalo stood placidly swishing away mosquitoes with its tail.

'How peaceful it is here,' the student said, smiling.

Jinda did not smile back. Squatting down by the fire, she stirred the soup. The gleam of the embers cast flickering shadows on the walls.

Should she add another handful of rice to the pot, Jinda wondered? But their storage bin was nearly empty, and every grain of rice precious. Maybe she should just add more water?

As she thought this over, Sri knelt beside her. She actually knelt, her knees tucked primly together, instead of squatting.

'I bought some rice with me,' the student said. 'And some salted fish. Shall I get it now?'

'Keep it,' Jinda said, without looking up.

'But it's meant for you. Ned said we'd be parasites if we didn't contribute our share of the food.'

Jinda wondered what 'parasites' were, but didn't want to ask. 'We don't need it,' she answered gruffly.

'I'll get it anyway.' Sri stumbled awkwardly to her feet, and rummaged in her knapsack.

Determined not to turn around and look, Jinda stared into the fire. She heard the sound of paper being rustled.

'Where should I put the rice?' Sri asked.

'Anywhere you want.'

The stranger circled the room, pausing by the jars of rock salt and dried chili on the shelves. Would she have the sense to pour it into the rice bin, Jinda wondered?

Then she heard the wooden cover of the bin sliding off: the stranger had found it after all.

'It's almost empty!' Sri called out, her voice reverberating in the hollowness of the bin.

Jinda flushed. 'We'll have plenty of rice after the harvest's over,' she said.

'Yes?' Sri sounded uncertain.

There was a pause, then Jinda heard the sound of rice

being poured into the empty bin, a swift constant drumming like the pattering of rain on a tin roof. Jinda yearned to look, but forced herself to remain by the fire. The stream of rice continued — how much did that stranger bring anyway?

At last Jinda could resist no longer. She jumped up and watched the flow of rice — such long, unbroken plump grains! Her mouth dropped open. It had been years since she had seen rice as beautiful as that!

The girl emptied the bag and, before Jinda could protest, crumpled it up and tossed it into the fire. Jinda watched it burst into flames — a thick brown bag without a single hole in it, wasted!

Sri knelt down by the fire, attracted by the burning paper. Then she peered into the pot. Beneath the pock-marked surface were some wilted leaves, and lemon grass stalks. 'What . . . what is it?' she asked.

'What do you think it is?' She gave the lemon grass soup such a fierce stir that a few drops splattered out onto the stranger's bare arm.

Sri took out a starched white handkerchief and wiped her arm with it.

So what if the ground nuts and tamarind leaves Pinit had gathered from the forest had given the soup a grey-green tinge, Jinda thought. Is it our fault that we've had two years of drought? Why should we be ashamed of what we have to eat? Still, Jinda wished bitterly that she could have cooked some chicken curry, or minced pork with basil. Then the visitor would have sniffed it hungrily and be eager for breakfast.

As it was, Sri was frowning slightly. 'Is it . . . is it for us?' she asked.

Burning with shame, Jinda pulled the pot off the fire and shoved it into a corner. 'No, it's for the pigs!' she cried. 'Didn't you see the big pig pen in our yard?'

Sri's smile was one of sheer relief. 'Yes, of course,' she said. Jinda stoked the fire without saying another word. Finally she filled another pot with water and scooped out five handfuls of the rice that Sri had brought, and poured it into the pot.

'Can I help?' Sri asked.

Can a rooster walk any faster with three legs, Jinda retorted silently. 'We might as well have your salted fish too,' she said, not looking at Sri. 'You can cut me a piece of it.'

Sri pulled out another paper bag and took out a dried fish from it. Holding it with two fingers, she turned it over dubiously. 'I . . . how . . . where do I cut it?' she asked.

Jinda stared at her in disbelief.

'I mean . . . along the front, or back? Or down the middle?' The girl's voice faltered.

Without a word, Jinda took the fish from her and started to cut it.

Sri watched helplessly. 'Ned says Thai students don't use their hands enough,' she stammered on. 'He says that's one of the biggest problem of Thai intellectuals today. All we've ever been taught to do is to think.' Sri sounded as if she was reciting from a textbook.

Sure, Jinda said to herself. Then how come you never thought to look at our pigsty and see that it's empty?

Chapter 4

The four students settled down to life in Maekung quietly, and just as quietly their presence was gradually accepted by the villagers there. Each morning they would set out with the families they were staying with, and join the harvesting in the fields. They had to be taught how to hold a sickle, how to grasp the rice stalks in one hand and slash through them with the other, but they learnt quickly, and seemed not to mind the hard work.

Despite her initial misgivings, Jinda grew to like, even to admire, the pale shy Sri. True, Sri worked very slowly, sawing clumsily with the sickle blade when a deft slash would have done. But Jinda saw how badly blistered the medical student's small soft hands became, yet Sri never said a word. Jinda would claim the place next to Sri in the row of harvesters, so that she could cut a wider swathe and help Sri with the work.

The November days remained cool and dry. The strong morning breeze would blow, swirling up eddies of dust in the stubbled fields. The day they were harvesting Lung Tong's field, the wind was especially strong.

A group of farmers were collecting the sheaves of rice,

tying them in huge bundles before hoisting them on to a shoulder pole. Jinda watched them as she slashed through her swathe of rice stalks. She had always liked the way the men moved, balancing the harvested stalks across their shoulders as they carried them over the fields to the threshing ground. But today she watched even more keenly. For the first time, Ned was among them.

Under the brim of her straw hat, Jinda watched him. Even with the weight of the rice stalks on his shoulder, he walked gracefully. He had learnt to walk with the long, rhythmic strides necessary to balance the load of rice stalks on the shoulder-pole. It looked effortless, but as he approached Jinda, she saw his forehead was beaded with sweat, and his shirt soaked through. It's not right, she thought, that a university student with hands so smooth and uncalloused should do such work.

She watched him as he passed, and smiled at him encouragingly. He looked at her, lost his footing, and stumbled. 'Careful!' Jinda cried. She sprang forward to steady the bundle of rice. For a second their hands touched, and then he straightened up and hurried on. Jinda stood looking after him, aware of the whispering around her.

Minutes later, two village youths carrying loads like Ned's walked by. One of them, Vichien, had often tried to attract Jinda's attention at wedding feasts or the temple fairs. As they approached her, Vichien pretended to totter backwards, flinging out his arms. The other one called out, 'Careful!' in shrill mimicry of Jinda. The girls nearby tittered loudly.

And Jinda felt as if hot coals were being poured down her neck. For the rest of the day, she was careful not to look at Ned again, much less speak to him.

After that incident, Jinda kept her distance from Ned, but she continued to grow closer to the medical student who was living in her house. After all, she thought, no one could gossip about her friendship with a girl.

And so, in the cool of twilight after work, she would wander down the quiet village paths with Sri, asking her questions about city life, and introducing her to the other families. Shy as Sri was with adults, she seemed quite at ease with children, and would play and laugh with them happily.

One evening, on one of these strolls with Jinda, Sri noticed that Nai Tong's youngest son had bloodshot eyes. His eyelids were puffy, and he kept rubbing them.

'What's the matter?' Sri asked him. 'Have you been crying?'

The boy shook his head vigorously. 'I don't cry,' he said.

'Are your eyes itchy then? They are? Can I have a look at them?'

Obediently the boy let Sri examine his eyes.

'Conjunctivitis,' Sri told Jinda. Then she turned back to the boy and said, 'Run home and ask your mother if I can put some eyedrops into them. It'll make the itch go away.'

Within a minute the boy was back, with a suspicious mother in tow. 'What do you want to do with Bui's eyes?' she demanded.

Shyly Sri mumbled a long explanation about infection and possible damage to the retinas. The mother listened grimly.

'How much?' Little Bui's mother asked, after Sri had stopped talking.

Sri blinked. 'Oh, it . . . it's free,' she stuttered.

'Go ahead, then,' Bui's mother said.

A crowd of curious children had gathered by the time Sri came back with a vial of clear liquid. Jinda held the boy's head as Sri squeezed two drops into his eyes. 'Come back tomorrow,' Sri told him. 'I'll have to do this for a few more days.'

The next day, Bui brought two other children with red eyes with him. 'Their mothers say you can do them too,' he announced, 'if it's still free.'

Soon it was not just conjunctivitis, but skin rashes, and stomach ailments and chest colds.

Each evening, when Sri and Jinda walked home from the fields, there would be a growing collection of children waiting for Sri underneath Jinda's house verandah, some with their mothers, some with older brothers or sisters. Sri always treated each child carefully, and never charged any money.

'Do you treat grown-ups, too?' Nai Wan's wife asked one night.

'I'll treat anybody who's sick,' Sri said.

'Not under my house, you won't,' Inthorn said. He had just bathed, and was trying to make his way through the crowd of children to the stairs. 'A man needs some peace and quiet in his own home. You start on the grown-ups, and I'll be having every farmer and his wife waiting under my porch.' He smiled at Sri. In the few days that the medical student had lived there, a relaxed, almost father-child affection had sprung up between them. 'I've been meaning to ask you, Sri, why don't you set up a table in the temple courtyard and treat your patients there?'

'A clinic,' Sri echoed, and her eyes shone. 'All those medical supplies I brought with me. I could start a real clinic!'

'And I can take my showers in peace!' Inthorn laughed.

Word about the clinic spread quickly through the village. The next night, before the bonfire was even lit in the temple courtyard, dozens of villagers were already gathered there.

Jinda felt rather uneasy when she helped Sri unpack her box of medicine on a wooden desk borrowed from the monks, and placed a kerosene lamp on it. 'You're all set,' she told Sri. 'I'll wait for you at home.'

'Jinda, don't go away,' Sri said, clutching her hand. 'You're so good with the children, and I . . . I don't understand their dialect sometimes. Please stay.' Jinda bit her lips. Much as she had grown to like Sri, she did not know how the other villagers might react to her joining this stranger in their midst. Her sister Dao had already warned her that Mau Chom, her father-in-law and the traditional healer, resented the strangers, and especially resented Sri, who he saw as an upstart rival to his own healing skills. Dao herself had pointedly stayed away from her own family ever since Sri started to live there. What, Jinda wondered, would her sister think of her helping at Sri's 'clinic'?

Sri's hand tightened on her own. 'Jinda, please,' she said. 'I'm scared.'

Jinda smiled. It was settled then, as simply as that. 'Fine, let's start,' she said.

The first person in line was Lung Teep, a wiry old man with ears that stuck out like little tree mushrooms on either side of his head. Lifting the cuff of his baggy trousers, he showed Sri a large, festering sore on his calf. Jinda held the lamp close as Sri opened the sore, drained the pus off, and applied some ointment on it. 'And take one of these pills before you sleep every night,' Sri said, handing him a little packet.

The old man took it with both hands. 'Thank you,' he said. Then added, 'Doctor.'

Sri and Jinda exchanged a quick smile

Many of the villagers who came had skin rashes, rough patches on their wrists and ankles where the fuzzy rice stalks had rubbed against their bare skin. There were also those with wracking coughs and high fevers. And countless others with intestinal problems. Quietly and efficiently, Sri applied ointment, gave injections, prescribed pills.

But a few came who Sri's medicine could not cure. A crippled young man, old women whose teeth were loose and rotten from chewing betel-nut, a girl with a harelip, all were told they could not be helped, and gently turned away.

One person whom Jinda felt particularly bad about turning away was Chart. He came late that evening, a clumsy young man who shuffled out of the shadows, his arms swinging loosely. As a boy, Chart was one of the few children to have studied at high school. He had once given Jinda a worn old geography textbook with pictures of children in different countries, which she still treasured. Then last year, while he was back in Maekung during a school holiday, he had been struck down with such a high fever that nobody thought he would live. He survived, but since then had been unable to talk coherently, much less read and write. Instead, he would wander through the village, playing with the smallest children until even they chased him off.

As Jinda explained all this to Sri, she saw the student frown. 'What does he want?' Sri asked.

Chart made a series of guttural sounds.

'He wants to go back to school,' Jinda interpreted for him. The boy ducked his head eagerly.

'So you had a high fever?' Sri asked. 'And you turned hot and cold, every few hours?'

Chart nodded, again and again. His thick fingers gripped the table.

Sri prised his fingers off, very gently, one by one. 'I can't help you, brother,' she said. 'You probably had malaria — the falciparum strain is still prevalent around here — and then brain damage caused by malarial fever.'

The boy looked at her, his eyes round and trusting.

'There's nothing I can do for you. Nothing, understand?' She had prised his fingers off the table, and now held his hand in hers. 'Nothing,' Sri said softly, blinking behind her thick glasses.

For a moment the boy stared at her blankly. He turned to Jinda, but she could only shake her head. Finally, eyes downcast, he stumbled off into the shadows.

Sri's hands were clenched into tight little fists. 'It's not my fault,' she said, her voice more terse than Jinda had ever heard it. 'How can I help people like him?' She pointed at the medicine in this cardboard box, and said, 'See that? It doesn't look like much, does it? A group of us, medical students, made the rounds of every pharmacy in Bangkok, begging free medicine off the owners. Night after night we'd do this, until we'd collected boxes and boxes of it. Filled half the storeroom at the Student Union office. It looked like an enormous supply. We felt we could heal every sick child in the country with it. Then we divided the medicine among each student group going out to the villages, and it didn't seem like so much anymore. Now,' she pushed the battered box away wearily, 'now it seems like nothing.'

Jinda quietly put her arm around Sri's bony shoulder. 'Let's go home,' Jinda said. 'You're tired.'

'Tired?' Sri echoed. She started putting some bottles back into the cardboard box, 'Yes, I guess I am. But I have no right to be.' Sri trimmed the wick in the kerosene lamp, and sighed. 'I have done so little yet. I have no right to be tired.'

Still Sri continued to administer her clinic. More people came, and because the lines got so long, and because whole families would come with the patient and then have nothing to do, Ned started small discussion groups around a bonfire in a corner of the temple courtyard.

And the farmers would gather every evening in growing numbers as word got around that some very interesting ideas were being talked about there. Before long, Ned's fireside discussions were as established a part of the village nightlife as Sri's clinic.

Invariably, the talk would revolve around the land rent. The harvesting was almost over, and the threshing already begun. Sometime within the next two weeks, the rent-collector would come in his truck, and in one day, haul off half the rice they had taken five months to grow.

As Jinda helped Sri apply ointments and bandages, she would listen to the discussion between Ned and the other farmers.

Ned had no solution as neat as Sri's bandages, but he would talk, late into the night, of other countries where the land-rent was only one-tenth the value of the crop, or of other villages in Thailand where farmers were beginning to refuse to pay this high rent.

One night, after the last of the patients for the evening had left, Jinda and Sri lingered by the kerosene lamp. A gust of night wind blew across the courtyard, tinkling the tiny bronze bells on the temple eaves. Around the campfire, only a handful of farmers talked with Ned. A few phrases — 'poor harvest', 'high rent', and 'already in debt with the money-lender' — drifted over to Jinda. Gradually, even their voices subsided, and in the stillness of the night, the whirr of cicadas grew louder.

Then, Jinda saw a lone woman emerge from the shadows of a big raintree, and walk towards her. Head draped in a homespun shawl, her face was in shadow. Only after she entered the circle of lamplight did Jinda realise who it was.

'Dao!' she exclaimed in joy. 'And you brought little Oi!' Jinda had hoped several times that her sister would bring her baby to see Sri, but every time she had even tried to suggest this, Dao would flare up, and say that the modern medicine Sri brought was just another form of Communism. Once Jinda had even heard Dao's father-in-law, Mau Chom, mumble a curse on the students, so she had stopped mentioning this to her sister.

Now Dao stopped short a few feet from the table, and hesitated. Her grandmother followed close behind, prodding her gently.

So that's who finally got her here, Jinda thought.

'Please, can I help you?' Sri asked quietly.

The old woman stepped up, her smooth thin shoulders gleaming in the lamplight. She took the habitual cigar from her mouth and tucked it behind her ear. 'It's little Oi,' she said, nodding towards the bundle in Dao's arms. 'He's sick.'

When Dao made no move to hand over the baby, the grandmother lifted him up from Dao's arms, and held him up to Sri.

With both hands, Sri carefully took the bundle. She unwrapped the threadbare cloth and looked at the large, unblinking eyes staring up at her.

'His head is burning, and he isn't even hungry anymore,' the grandmother said.

Dao spoke up then, her voice as soft as the night wind. 'He doesn't want rice-gruel, nor will he try to breastfeed now,' she said softly. 'And he doesn't even smile when I sing to him anymore.'

Sri's long pale hands moved up and down over the baby, probing, stroking, examining. She stopped at a large sore on his right arm, and frowned. 'What's this?' she asked.

'Just a mosquito bite,' Dao said defensively. 'Everyone has them. He's had several before, but this one doesn't seem to heal.'

Sri said nothing, her face grim. Telling Jinda to hold the lamp up, she peered into the baby's ears, pressing at his temples. He whimpered.

'Left ear's infected, too,' she said. 'That's probably what's making him feverish.' Her hands strayed over the baby's bloated stomach, up his ribs and to his thin stem of a neck. 'Kwashiorkor,' she said. 'Marasmic Kwashiorkor. So malnourished he doesn't have the strength to fight a simple infection anymore.' She looked up at Dao. 'What do you feed him?' she asked.

Dao chewed at her underlip. 'I . . . I do my best,' she faltered. 'Rice porridge mostly, and some mashed bananas when we have any . . .'

'Any milk?'

Dao turned her face away. 'My milk ran dry weeks ago,' she murmured, her voice low with shame.

'But your baby needs protein, and . . .'

'Oh give him some then, please!' Dao stretched her hands towards the bottles of medicine lined up neatly on the table.

'But I can't, it's . . .'

'Please, just a little?'

'You don't understand,' Sri said, biting her lips. 'Protein isn't medicine. It's just meat, fish, milk . . .'

'Meat?' Dao echoed, bewildered. 'And fish?'

The two women stared at each other for a long moment. Finally the grandmother spoke.

'We have no meat, no fish, no milk,' she said kindly. She relit her cigar with a twig dipped into the lamp, and took a long puff from it. 'When the paddies are knee-deep in rainwater, we can catch snails, or soft-shelled crabs, and even some catfish.' She blew out a stream of cigar smoke and continued dreamily. 'Or in the winter, we can go into the forest, and find pink and white mushrooms under moist ferns, or we cut the spikes of newly sprouted bamboo shoots from the groves.' She tapped the ash off her cigar lightly. 'But what is there now when the fields are cracked and hard, and the mountainsides barren even of leaves? Don't talk to us of meat and fish, child. We're hungry, all of us — but the little ones most of all.'

Sri took off her glasses and rubbed her eyes wearily. 'Of course,' she whispered. 'I should have realised.'

'Those eye-rings of yours,' the grandmother smiled, 'They're probably very good for reading books, but are they as good for telling you what's right in front of you?' She leaned forward, and stroked the baby's cheek with a rough, wrinkled fingertip. 'This is my first great-grandchild,' she said quietly. 'Can't you help him at all?'

Sri looked very tired. 'Medicine can cure the sickness he has,' she said, 'but it won't help him grow to be a healthy boy.' She held the baby out to Dao, but Dao stood with her hands stubbornly by her sides, refusing to take him back.

'Please,' she said, looking at a bottle of bright blue pills. 'Give him some medicine, then. Just a handful. He doesn't need much. He's just a baby.'

'I can give him a dose of antibiotics and some vitamins,' Sri said quietly, 'but it isn't food.'

Dao nodded, and took her baby back, carefully wrapping

him in the shawl again. She accepted the package of pills from Sri with bowed head. The she and the old grandmother both slipped back into the shadows again.

Jinda lowered the wick of the kerosene lamp so that there was only a little circle of light around them now. Sri put her glasses back on, and started to pack up the medicine.

'Neomycin, aureomycin, ethromycin,' she chanted softly, reading off the labels of each bottle before putting them away in the box. 'Tetracyclin, hydrocortisone, teramycin . . . all this medicine.' She put the last bottle in and closed the flaps of the box. 'All this medicine, and what use is it?'

The night was very still, and the stars and fireflies had struck up their mute dialogue. Somewhere a lone dog howled in the night.

'Can doctors heal hunger, Jinda?' she asked . 'Will any amount of penicillin cure poverty?'

Jinda shook her head. She had not understood much of what Sri was saying, but in her heart she knew this much: little Oi was not going to get well. She felt angry and hopeless, but when she saw the bewilderment in Sri's eyes, she wanted only to comfort her friend.

'You've helped quite a few farmers here already,' she said.

'It's not the few farmers here and there who are sick,' Sri said, as if talking to herself. 'It's the whole society. It's our country that's sick. What then can I do? What use am I?'

Jinda picked up the box of medicine. it felt heavy and solid in her arms. 'You've been a great help,' she said. 'You've cured so many people in Maekung already. You've done a lot.'

Sri shook her head. 'Not enough,' she said. 'Not nearly enough.'

Chapter 5

Three days later, Dao's baby died.

There was no funeral, no monks came to pray, no special meal was cooked for the few mourners. His little body was simply placed in a rough coffin, where he looked even more fragile and shrunken than before.

He was cremated near the temple grounds, in a clearing at the centre of a grove of large Bodhi trees. His ashes, like those of countless others before him, were scattered by the wind to the rice fields beyond.

Even after the cremation fire had died down, and the embers were burning low, Dao remained in the shady grove, sitting against the trunk of a gnarled Bodhi tree. Sri and Jinda and her grandmother, together with a few of the village women, sat nearby to keep her company.

A dragonfly glided down and alighted on Dao's arm. Sri reached over and brushed it away. 'I'm very sorry about little Oi,' Sri murmured, her hand on Dao's arm.

Dao flung Sri's hand off. 'Saying sorry isn't enough,' she said grimly, 'for killing my child.'

Jinda stared. Was that hard, tight voice really her sister's?

It had sounded so strange, perhaps because she detected Mau Chom's spite in it.

'I did all I could to help him,' Sri stammered. 'Surely you know that?'

'You didn't help him. You killed him, and you know it!' Dao said, in the same brittle voice.

'But he . . . he was already dying, Dao,' Sri said shakily. 'I told you that. I said he wasn't being fed enough meat and milk . . .'

Dao seized on the last word. 'Milk, is it? You know my milk ran dry weeks ago. Are you blaming me then?'

'Of course not,' the student said. 'If anything's to blame, it's this system of paying such a high rent to . . .'

But Dao was not listening. She got up, holding onto the tree for support. There was a dazed look on her face. 'No, I am to blame, after all. I had no milk.'

'Dao, you couldn't help that,' Jinda said. 'Come, let's go home, you're tired.'

'Little Oi wanted milk, so badly. And I had none to give him,' Dao said. The checkered shawl in which she had wrapped her baby was in her hands. She was twisting it into a tight coil. 'If I had milk, he wouldn't have died.' Her voice broke. Flinging the shawl down, she ran out of the Bodhi grove, and was gone. Her bare feet kicked up tiny swirls of dust as she stumbled up the path to the mountain.

Jinda watched her, not knowing what to do. She felt someone shake her shoulder.

'Go after her, child,' her grandmother said quietly. 'Your sister is without mother, and now without child. You're the one closest to her now.' Her eyes were opaque and misty, as much from age as from sorrow. 'Go and talk to her, child,' she said.

So Jinda got up and walked out of the Bodhi grove. The sun was fierce, and the sand was hot between her toes as she followed Dao's footsteps up the hillside.

She found Dao sitting on a log under a tree whose vines

40

they used to swing on years ago. Dao looked up as Jinda sat down beside her, but said nothing.

Tongue-tied, Jinda gazed at the view below them. In the dull glare of the afternoon sun, everything looked dead. The village was one flat, relentless shade of brown. Brown dust on brown roads. Brown thatched roofs over cracked walls bleached a dull grey-brown. Brown hay stacked on dusty brown lofts. Brown leaves drooping from gnarled brown trees. Everything was brown, brown, brown. If death has a colour, Jinda thought, it is brown.

'Dao,' Jinda said awkwardly, 'I'm sorry Oi is dead.'

Her sister did not reply.

'Please don't be sad,' Jinda said. 'You can always have another baby, can't you?'

Dao did not move. 'Shut up,' she said.

'But it's not so hard, is it? To have another baby?' Jinda ploughed on, ripping open her sister's reserves of pain without realising it. 'It only takes a few months, and you said before you even liked being pregnant . . .'

'Shut up!'

But Jinda did not know how to stop. 'But you said you liked feeling him kick inside you, remember? You let me feel him, my hand on your belly. What a strong kick he had! Did it hurt, Dao? What did it feel like?'

'It felt all right,' Dao said grudgingly, after a long pause. 'It felt like he was trying to touch me.'

'And when he was born, did that hurt? They wouldn't let me see, you know. Was there a lot of blood?'

'It hurt,' Dao said, but with a touch of pride. 'He was a big baby when he came out. With fat hands and little soft ears — just like his father.' Dao talked then, her words coming slow and thoughtful, as if each one was precious, and must be savoured. 'I took him out and bathed him in a bucket by the well at first, do you remember?' Memory upon memory she dredged up and laid before Jinda: how she had suckled him at her breast, how he had first smiled at her, how he had held onto her thumb with both hands. As she spoke, her voice became lighter, and once she even smiled.

Jinda urged her on, wanting her to smile again. 'And that time we took him bathing in the river,' Jinda prompted, 'on the first day of the harvest. He reached up and stroked your cheek, like this . . .' she leaned over and placed her hand on Dao's face, hesitantly.

Dao held onto her sister's wrist, and pressed her face into Jinda's palm. She's crying, Jinda thought, my big sister never cries. But the tears came, cool and swift and silent, trickling through Jinda's fingers, and down her arm.

They sat there on the log for a long while, watching the afternoon shadows lengthen in the valley below them. Neither of them spoke much. Towards twilight they heard the familiar jingle of cowbells, faint and musical in the distance. They knew it was their little brother bringing the buffalo home, and so they got up and went down the hill to meet him.

In some subtle but pervasive way, little Oi's death affected the entire village. Around their weaving looms, women would cluck over how thin their own children were getting. And by the temple bonfire at night, farmers would brood over how little rice they'd have left this year, after paying the rent. And in all their conversations would lurk the unspoken question: whose child would die next? In this way, the baby's death became more than Dao's personal sorrow; it was now a matter of common concern for the whole village.

Inextricably tied to this was the villagers' anxiety about the rent. It was already early December, and the threshing almost finished. The mount of rice grain in the middle of the threshing ground was growing day by day, and soon Dusit, the rent-collector, would be due to take half of it away. Resented even during the best of harvests, Dusit's arrival this year was especially dreaded. After he had collected his rent, would there, the villagers asked each other, would there be enough rice left to last them for the coming year?

Everywhere around her, Jinda heard farmers talking about the high rent. Often, she noticed, Ned or Jongrak would be in the midst of the discussion, quiet but attentive. Jinda would stay on the fringe of these groups, listening with keen interest, but not saying anything.

One afternoon, however, she had the chance to talk to Ned directly about this. She was sitting by the river bank, watching her little brother splash river water on his buffalo.

It was peaceful there. A magpie perched on the buffalo's back, pecking at the insects on it, as it swished its tail back and forth in a graceful arc.

A dry twig snapped behind her and, glancing around, Jinda saw Ned walking down the trail towards them, a cloth towel draped around his bare chest.

'I see I'll have to share my bath water with your buffalo,' he said, smiling at Jinda.

Jinda tried to hide her pleasure at seeing him. It had seemed a long time since she had had the chance to talk with him. 'You could use the water upstream of him,' she said lightly, not sure whether she should leave now that he had announced his intention of bathing there.

'Oh well, I'll just help your brother wash him for a start. How about it, Pinit?' Ned called out.

The little boy grinned. 'Fine!' he said. 'Come on in.'

Jinda watched as Ned and Pinit scooped handfuls of water over the buffalo's legs, rubbing off the patches of mud caked there. The animal stood quietly between them, its eyes dreamy and unfocused.

'So how are things at home?' Ned asked conversationally.

Pinit shrugged. 'Nobody talks much anymore. Father worries about the rent-collector coming all the time. I can't even make him laugh now.'

'Paying half the harvest as rent is far too high!' Ned said, loud enough so that she knew he was actually addressing her rather than Pinit. 'Even after a good harvest, it's too high. But in a drought like this, it's a crime! How can you give so much away?'

'We don't give it away,' Jinda replied drily, 'it's taken from us.'

'You don't have to let them take it away,' Ned said. 'Keep a bigger share for yourself.'

'How?'

'Resist the rent. Keep two-thirds of your rice for yourselves, and give one-third to this Dusit. That's only fair.'

'That's easy for you to say,' Jinda retorted. 'It can't be done.'

'Why not?'

'It's against the law, that's why not! We'll be arrested.'

'But it's not against the law,' Ned argued.

'Well, it might as well be. It's always been this way.'

'But that's not the law. That's only tradition.'

'We'd be forced off the land,' Jinda said. 'Replaced by other tenant farmers who aren't troublemakers.'

'Not if all Maekung decides to resist together,' Ned argued. 'In some countries the rent is only one-third, or even one-tenth of the crop.'

'Other countries!' Jinda felt a surge of impatience. 'We are not farming in other countries.'

'But we can make Thailand like other countries,' Ned argued. 'Our new government is considering a law limiting the amount of rent due to a landlord. If this legislation is approved . . .'

'You talk like a book,' Jinda snapped. 'You treat our lives as if they were some new exercise in a schoolbook. You don't know how things are really like in a village.'

For a long moment Ned looked at her without speaking. Then he waded back to the riverbank and sat down next to her. 'I do know,' he said quietly. 'I grew up in a village even poorer than this one.' Awkwardly, with none of his usual fluency, Ned started to talk about his childhood. During the bad years, he said, after drought or locusts or rats, there would be very little rice left after the rent was collected. 'One year my father went away to find work in Bangkok,' Ned said. 'I never saw him again. The same year my mother died in childbirth. That's right, like yours.' His hands were tightly clenched, and Jinda suddenly wanted to smooth open his fists, and put her palms over his.

'Go on,' she said.

'There isn't much more,' Ned said. 'I was sent to live as a temple-boy in a local monastery, where I stayed until I was sixteen. I was a pretty good student — there wasn't much else to do there — and got a scholarship to study in a Bangkok high school, and then at Thammasart.' He looked up at Jinda, and smiled ruefully. 'I'm sorry if I sound like a book, Jinda,' he said quietly, 'but sometimes it's easier to talk like a book than like a person.'

Impulsively, Jinda reached over and covered his fist with her hand. He unclenched his hand and held hers, gently.

'Father!' Pinit suddenly shouted. 'What're you doing here?'

Startled, Jinda and Ned sprang apart, as Inthorn rushed down the trail at them. He looked furious.

'There you are!' he said. But his scowl was directed at little Pinit, and he barely seemed to notice Ned or his daughter on the riverbank. 'I thought you'd be down here with that . . . that damn buffalo of yours!'

'Oh Father,' Pinit said weakly. 'I was going to tell you about it, really I was. I just wanted to wash him first . . .'

Jinda stood up, discreetly placing more distance between herself and Ned, and asked, 'Tell him what, Pinit? What happened?'

'Yes, son,' Inthorn said sternly. 'What did happen? Let's hear your side of it.'

Pinit hung his head, staring fixedly at his buffalo. 'I was racing cousin Daeng home from the fields this afternoon,'

Pinit began in a small voice. 'And winning too. But I couldn't make my buffalo stop. It charged straight through Lung Tong's fence, and nearly threw me off.' He stroked the buffalo's forehead and added with a hint of pride, 'It finally took four men to restrain him.' The little boy looked up at Inthorn with big, sad eyes. 'Father,' he asked, 'You . . . you're not going to drill his nose, are you?'

Inthorn nodded grimly. 'This isn't the first time the animal has been unruly, son. It needs to be tamed.'

Pinit swallowed hard. 'It's going to hurt him, isn't it?'

'No.' Inthorn said more gently. 'Not much.'

He's lying, Jinda said to herself. She thought of the nose-drillings she had seen, with the poor animals snorting with pain and fright as the wooden drill pierced through their nostrils. All this so that once a hemp rope was threaded through the tender nostrils, the animal would obey every tug from its masters. Looking at Pinit's forlorn little face, Jinda felt sorry for both her brother and the young buffalo.

Ned must have felt a similar stab of sympathy for Pinit, because he strode over to the boy and patted his shoulder. 'It won't be bad,' he said. 'We'll polish the wooden drill so smooth and sharp your buffalo will hardly feel a thing. Come on.'

He and Jinda exchanged a quick smile as Pinit immediately brightened up and asked his father for some sandpaper.

On the way home, Ned and Inthorn started to talk of the rent again, as Jinda walked a few steps behind. Little Pinit was still at the river, giving his buffalo a final scrubbing.

'Have you decided what to do yet, sir?' Ned asked the farmer. 'Are you going to resist paying the full half of the rice?' Jinda knew that Ned was especially interested in Inthorn's decision, since it would influence those of the other farmers in Maekung.

'I don't know,' Inthorn sounded wary. 'Any more news of

the other villages in Chiengrai or Pitsanuloke provinces about rent resistance?' he asked.

'Just what was on the radio last night,' Ned said. 'More and more villages are holding back two-thirds of their rice crop. The movement is spreading, sir.'

'Yes, and more and more village leaders have been arrested, or even killed.'

'Rumours . . .' Ned began.

'Nai Pruk of Batong was gunned down last week,' Inthorn said, and Jinda could hear the pain in her father's voice. 'He was my cousin.'

'One man,' Ned said.

'I could name you a dozen others, but I'm sure you know of them as well as I do.'

'Some things are not easily won,' Ned said quietly.

'And some things,' the farmer answered, 'are not worth fighting for. It's fine to talk of fighting and winning, Ned, but what if we lose? We don't have much now, but if we fight and lose, won't we stand to lose even what little we have?'

'But even Pinit's buffalo has fighting spirit. How much more fighting spirit a man should have!'

Jinda hung back, and let the two men walk on ahead of her. It was dusk, and the shadows around her were lengthening. In the distance the sun-washed mountain ridge curved in a wide arc, encircling their valley. It was the most peaceful time of her day, and Jinda wanted to hear no more talk of fights and killings.

Pinit came up behind her, astride his gleamingly clean buffalo. Jinda reached out and stroked the animal's rough, grey hide. 'When we get home,' she told her brother, 'I'll give you a big piece of rock salt for the buffalo to lick.' Little Pinit grinned with delight, and Jinda felt a little better.

Chapter 6

Early the next morning, several curious villagers gathered by the gate of Inthorn's house to wait for the nose-drilling. Nai Tong and Ned, among others were waiting by the mango tree, as Pinit led his young buffalo towards the tree.

'Do you really have to do this, Father?' Pinit asked, as the farmer slipped a muzzle over the buffalo.

'I'm afraid so, child. It's the only way to tame it. A buffalo's got to learn to serve its master its whole life.'

'Unlike a man,' Ned said gravely, glancing over little Pinit's head at Inthorn. 'A man should be his own master, right, Pinit?'

Inthorn glanced at Ned sharply, but said nothing. Grimly, he tied a rope to one end of a heavy bamboo pole and stuck the other end deep into the soil at the base of the mango tree. As he led the buffalo into the space between the tree and the pole, Nai Tong pulled the rope tight around the tree trunk, until the buffalo's head was held fast.

The buffalo strained against the pole. Its eyes bulged, and it breathed so hard that white froth collected on its muzzle.

Without ceremony, Inthorn put one hand on the buffalo's muzzle, then started pushing the drill into one of its nostrils.

Pinit whimpered.

The animal pawed the ground furiously, its eyes terrified. It tried to jerk free, but its head was held fast by the tree trunk and the bamboo pole. Blood trickled out of its nostril, staining the sand.

Pinit was crying now, great round teardrops zigzagging down his cheeks. Inthorn glanced at him, then, wincing, plunged the drill through the wall between the nostrils.

Just at that moment, Pinit wailed.

Startled, Inthorn suddenly jabbed down with the wooden drill with such force, that it pierced the nostril wall, shot out of the nostril and plunged into Inthorn's other hand.

There was a scream — Jinda? Her father's? — and Inthorn doubled over, clutching at his hand. Jinda caught a glimpse of the wooden drill as it dropped beside him. It was stained a deep red.

She ran over to her father, and tried to support him. To her horror, she saw that his hand was pierced through, and his thumb hung, half-loose, only by the bone. Inthorn was staring at it, his breath coming in short pants.

Jinda dropped down beside her father, who slumped against her for support.

'Get Sri,' she gasped. 'Hurry!'

'No, get Mau Chom!' someone in the crowd — was it Dao? — shouted.

After what seemed an eternity, Jinda saw the crowd parting for someone. Sri, Jinda thought with relief.

It was not the young medical student who appeared, but a plump old man in a dirty singlet pushing his way through the crowd. With a sinking feeling in her stomach, she recognised Mau Chom.

Inthorn was lying half propped by the mango tree, and half supported by Jinda. Very reluctantly, she held out her father's hand for the spirit doctor to examine. The bleeding had eased somewhat, but the gash was jagged and deep.

Mau Chom looked at the cut in distaste. Then, sitting

back on his heels, he muttered a low chant over the thumb, rocking himself rhythmically. At the same time, he uncorked a bottle he had brought with him. The strong smell of home-brewed rice wine wafted over, and he sniffed it appreciatively. Then, with a flourish, he tilted his head far back, and poured the whisky into his mouth. There was a brief gurgle as the liquid rushed through the bottleneck. He gulped several times, then leaned over and spat the last mouthful over the cut hand.

Jinda turned away; she felt sick. A mixture of brownish wine saliva, and blood dribbled down her father's hand.

'Carry him home. Put him near the family altar, and make sure his head points east,' Mau Chom said, with pompous authority. 'I'll check on him tomorrow.'

News of their headman's accident spread quickly throughout the village, and neighbours streamed in to visit, bringing medicinal herbs or broth. Like moths around an oil-lamp, they stayed to hover around him. They were waiting, Jinda realised, not just for news of his condition, but for his advice. After all, Dusit was rumoured to be coming to collect the rent in a few days and they still hadn't decided whether or not to resist paying the full rent. And so they gathered, talking among themselves in hushed tones, waiting for their leader to get better.

But Inthorn did not get better. His fever continued to climb, and that night he was completely delirious.

Jinda kept watch anxiously. Dao was there too, whether on Mau Chom's instructions or on her own initiative Jinda did not know.

Several times during the night, the farmer flung off his thin blanket, and struggled to sit up. Jinda and Dao tried to restrain him, but he only struggled all the more. For hours, he tossed about, begging some invisible demons not to drill his nose. The moist towels which his daughters applied to his burning forehead did not seem to calm him down at all.

'He's getting worse,' Jinda whispered shakily. 'What will we do?'

'Get Mau Chom to say another chant over him,' Dao said.

Jinda frowned. 'The first one hasn't done much good,' she answered. 'Let's have Sri look at him instead. She's got medicine, and . . .'

'No!' Dao said fiercely. 'Mau Chom said if anyone else tries to heal Father, his spell won't work. Especially Sri. She's not a doctor, she's a witch. She killed my little Oi. You want her to kill Father too?'

'But your baby was desperately sick. There's still hope for Father if we ask Sri to . . .'

'No!'

'But Sri has helped a lot of people get well. Can't we . . .?'

'No!' Dao said, and she sounded so vehement, that Jinda did not protest again.

The next day, Mau Chom arrived and, without even taking his shoes off at the door, came in. He glanced around at the room, and at the still feverish Inthorn moaning in the corner. Fanning himself with a sheaf of ricepaper painted over with hex signs, he nodded approvingly.

'Windows shut good and tight,' he observed, 'and he's pointed in the right direction.'

'But his fever's climbing,' Dao whispered.

'That's to be expected. Evil spirits are battling with the good ones. Makes for a lot of heat. I'll say my most powerful chant for him. It's his only hope.'

Jinda frowned, and knelt beside her father, protectively.

Mau Chom sucked in his paunch, knelt down beside Inthorn's head, and passed his hands over the farmer several times. He began to chant in a low, monotonous voice.

In the middle of the ritual, Jinda noticed that Sri had crept in and was watching the spirit doctor intently.

When Mau Chom finished, Dao thanked him. 'Will he get better now?' she asked.

'Hard to say. I may have to say another chant or two.'

'And then . . . he'll get well?' Dao asked nervously.

Mau Chom shrugged. 'Sometimes evil spirits win, despite my spells. I can't heal everyone, you know.'

I know, Jinda said silently. I remember how Mother died, wide-eyed and burning with fever, after Pinit was born. Your fancy chants hadn't worked then, either, Mau Chom. Aloud she said, 'Maybe we could have Sri look at Father. I mean, she's a doctor too. It can't hurt to have . . .'

Mau Chom swung around and glared at Jinda. 'How dare you compare that little fraud to me?' he demanded. 'She's no doctor!'

From the doorway Sri suddenly spoke up. 'True, I'm only a medical student now,' she said quietly. 'But in another year I'll be a doctor.'

'You, a doctor!' Mau Chom laughed, and turned to Dao. 'You let her touch my patient, and my spells will be broken, do you understand?'

'Your spells can't be much good if they can be broken so easily,' Sri shot back. 'In fact, you've done more harm than good already, spitting that mouthful of germs onto the wound.'

'Germs?' Mau Chom glowered at her. 'It was sacred rice-liquor.'

'All right, the alcohol is a disinfectant, but your bacteria counteracted that.'

'Go ahead, little miss witch-doctor. Roll your magic words out,' the older man sneered. 'But if you so much as touch one hair on this man's head, he'll die.' With that, he got up and stomped out of the room.

The three girls looked at each other in silence. Then Dao said. 'You heard him, Sri. Go away.'

'But your father's hand is badly infected. He should have some antibiotics, and a tetanus shot . . .'

'He doesn't need your medicine.'

Sri turned to Jinda. 'Talk to your sister,' she begged.

'Make her listen. You've worked with me at the clinic. You know how many people I've helped.'

'I know,' Dao snapped. 'I know how you 'helped' my little Oi.'

'Your baby never had a chance, Dao. He was severely malnourished to begin with. But there's still hope for your father . . .'

'I said, go away.' Dao said, her voice hard.

Sri shook her head helplessly. Then, without another word, she left.

That night, Inthorn's fever mounted. He shivered violently, and stared glassy-eyed at his family, without recognition. Jinda and Dao continued to take turns watching over him.

After the second temple gong, Jinda relieved an exhausted Dao, and took up her vigil over her father. What if he didn't get well, she thought? She had seen her mother die a few years ago, trembling feverishly much like her father now. Would Sri's medicine have cured her, when Mau Chom's spells proved useless? Shouldn't her father be given the chance that was denied her mother, of being treated with modern medicine?

Then she noticed a light flickering in the doorway. She looked up, and there was Sri. Her glasses glinted in the light of the oil lamp she was holding.

'Please, Jinda, may I come in?' she whispered.

For a moment Jinda hesitated. She looked over at the far side of the room, where Dao was now sleeping. Taking a deep breath, Jinda motioned for Sri to come in.

The medical student crept into the room and knelt down beside Inthorn. She reached for the farmer's bandaged hand, and turned it over. Carefully, she started peeling the rag away. The pus had oozed out, dried in a thick yellow crust around his cut, and stuck to the cloth. The thumb was horribly swollen by now, the whole palm red and puffy. As Sri ripped the last shred of bandage away, she tore the crust, and a stream of pus trickled out. It smelled putrid.

'Good Lord,' Sri breathed. With gentle fingers, she probed his arm, his wrist, and finally his thumb. Inthorn moaned and tried feebly to twist away. 'It's pretty bad,' she said without looking up. 'The infection's spreading. It might even develop into gangrene if it gets into the bloodstream . . .'

'Can you cure him?' Jinda asked.

Sri looked up then, her face grim in the moonlight. 'I can try,' she said.

Jinda nodded. 'Try,' she said.

Sri flashed one of her rare smiles. 'Thank you, Jinda! Now, get me some water — boiled water. And I need more light.'

Jinda lit an oil lamp, turning the wick up high, and put the kettle on.

Together they knelt over the farmer. As Jinda held his hand firm, Sri washed it clean with warm water, swabbing at it with white cotton. It took a long time, but at last the blood and pus was gone. Now they could see how the cut had sealed over, and a taut mound of pale yellow had built up under the skin.

Sri drew out a thin knife and held it over the flame. 'Now. Hold on to his hand. Tight,' Sri muttered. When, with quick precision, she cut along the original tear. Immediately a ribbon of pus spurted out, and following that, a trickle of blood. Sri heaved a sigh of relief. 'Hey, you can let go of his hand now,' she said softly.

Only then did Jinda realise that she was still gripping her father's hand with all her strength, even though Inthorn had done nothing more than tug away weakly.

Sri washed the hand clean, then poured some clear liquid onto the wound. White foam erupted from the cut. 'What's that?' Jinda asked in wonder.

'Just some hydrogen peroxide,' Sri said. 'It's a cleaning

solution.' She kept squeezing more of the liquid on, until the foaming stopped.

The cut was revealed starkly now. Deep and jagged, it stretched from one side of Inthorn's thumb clear to the other. Only the bone held the thumb onto the hand. Sri examined the wound carefully, then took out a small packet of black thread and a curved needle. 'I'd give anything for a little anaesthetic,' she said.

Jinda held her father's thumb in place as Sri proceeded to stitch along the cut. Each time her hooked needle jabbed through, Inthorn flinched violently, but he was too weak to twist away. Sri was sweating now, small beads running down her cheek to collect on her chin. She was careful to make sure they dropped away from the wound.

There were eleven stitches in all, running crookedly across the farmer's hand. 'I wouldn't have made a very good seamstress,' Sri said wryly. She applied some white cream over the cut, then wrapped the whole hand in gauze.

Inthorn's moans had subsided, but his face was drenched with sweat, and the pillow under his head was soaked through. He mumbled inarticulately, then fell into a deep sleep.

'He'll rest better now,' Sri said. 'Give him two of these pills every four hours. And let him sleep as much as he wants.'

Jinda rolled down the wick in the lamp, and realised that her hands were trembling violently. She smiled shakily at Sri. 'Thank you,' she said.

'Thank YOU,' Sri said, 'for trusting me.'

After Sri left, Jinda sat watching her father for a long time. He slept peacefully now, his breathing even and regular.

A night owl hooted in the distance, and a passing breeze rustled the palm fronds out in the yard. Jinda felt a twinge of sadness. If only Sri had been here when Mother was sick, Jinda thought, things might have been so different. Then she shooked her head. It was too late for that now. At least her father would get well, thanks to these strange students and what they brought to Maekung.

At dawn, just after the fifth temple gong, Inthorn stirred. In a hoarse whisper, he asked for some water. Jinda lifted a ladle of well water to his lips, and slipped in the two pills that Sri had prescribed. Still dazed, he hardly noticed swallowing the pills, and drifted back to sleep.

He slept for most of that day Jinda was quick to suggest to Dao that she return to her father-in-law's home, and tell Mau Chom that 'his' patient had responded to his chants, and that he needn't come again.

By the end of the afternoon Inthorn was restless — and hungry. 'What's for dinner?' he asked weakly. And Jinda, barely able to contain her tears, rushed to the little kitchen and prepared a special meal for him.

Within minutes she was back, bearing a tray of food so carefully arranged that anyone would think she was offering delicacies to the monks. There was lemon grass soup, and a sliver of salted fish over a mound of rice, and a bowl of boiled peanuts which the neighbours had brought. As an after-thought, she had even added a hibiscus in full bloom.

'Here, Father,' Jinda said, placing the tray on a low round table in front of him. 'You haven't eaten for days, you know.'

Inthorn wetted his lips. He reached for the peanuts, then grimaced with pain. For the first time, he noticed the white gauze bandage swathing his thumb. 'I hurt myself . . . how?' he asked.

'Pinit's buffalo, Father,' Jinda said. 'You were drilling its nose when . . .'

'Drilling its nose,' Inthorn repeated slowly, as if he were waking up from a pre-dawn dream and was trying to remember the details. 'So it really happened. I thought somehow it was my nose . . .' He looked at his daughter and smiled. 'What strange dreams I've had, Jinda.'

Jinda smiled back, thrilled just because he had recognised her and called her by name.

'And all this white swathing?' he asked, studying his hand. 'Who did this? I remember Mau Chom . . .'

'No, Father, Mau Chom didn't do this. He only chanted spells.'

'Who did this, then?'

'Sri did, Father.'

'Sri?' Inthorn frowned. 'She's just a young girl. Is her magic really stronger than Mau Chom's?'

'You've seen her work at the clinic, Father. She's good.'

'Yes, but that was only for small ailments. I always thought Mau Chom was more powerful for sicknesses of life and death.'

'Apparently not, Father,' Jinda said drily. 'Sri's magic is powerful too. It must be, she cured you when Mau Chom couldn't.' Jinda took out the little packet Sri had left with her, and poured the capsules out into her hand. White and red, they gleamed like wet pebbles in her palm. 'See, Father? She gave you this medicine for the fever. You've taken six already, and it's time to take two more. Go on, take two. It's worked so far.'

The farmer looked at the capsules, frowning. 'What is it?' he asked.

Tetracyclin. Jinda could remember the lilting name Sri had used for those capsules during the many times she had prescribed them at the clinic. 'Tetracycline,' she said softly to her father, then smiled. 'Magic,' she amended.

Her father smiled back at her, then meekly picked up two capsules and swallowed them.

For the next few days, even though Inthorn was past danger, Jinda guarded him fiercely, anxious that he should rest in quiet. Dao hovered around, offering to bring Mau Chom to say another chant, until Inthorn grew irritable, and Jinda hinted discreetly that her father-in-law probably wanted her back in his house.

It was not so easy, however, to keep the growing numbers of Maekung farmers at bay. As word got around that Inthorn was recovering, more and more villagers gathered in the clearing under the verandah, and waited patiently to see their headman.

Jinda, backed by her grandmother, at first allowed only Inthorn's closest friends up to see him. Lung Teep, Nai Tong and Sakorn would climb up the stairs quietly, and sit in a little row outside the farmer's bedroom doorway, carrying on a hushed conversation with him, as the others strained to listen downstairs. They desperately wanted to know if Inthorn, their headman, would keep two-thirds of the rice from the fields he himself rented, and whether they should follow his example.

After three days of being confined at home, mostly in his bed, Inthorn grew increasingly restless. He wanted to see how the threshing was getting on, he told Jinda. He needed to be in the fields with the rest of the villagers.

Still Jinda resisted, until one afternoon, as she was drawing water from the well, she saw her father walking, alone and very slowly, down the village lane towards the fields. Her first impulse was to run after him and stop him, but something about the resolute set of his shoulders stopped her.

As he walked on, Inthorn was quickly joined by neighbours and friends, until, by the time he reached the threshing ground at the edge of his field, a small crowd had gathered around him.

Jinda set down her bucket of well-water, and hurried after him. Dozens of people were already there, Ned and Sri among them, and, as she approached, Mau Chom and Dao came up as well. This was the first time Inthorn had set foot outside his house since his accident seven days ago, and the villagers were obviously curious to see what he was about to do.

Slowly, Inthorn walked to the threshing ground in his two fields, and squatted down by the mound of rice grain there. As if almost oblivious of the people pressed around him, Inthorn scooped up a handful of rice, and let the unhusked, brown grain trickle through his fingers.

'For almost fifty years now,' he said dreamily, to no one in particular, 'I have grown rice in these fields, and for fifty years, seen half of what I have grown carried off.'

He looked up at the fields that stretched out before him, dry and stubbled in the afternoon sun. 'I know every inch of those fields. I have ploughed them and planted them, weeded them and fertilized them, harvested them and reaped the rice grain from them. I think of them as mine.' His eyes lost their dreamy look, and he scanned the faces of the people around him. 'Brother Teep,' he addressed his closest friend. 'Your eight rais of land, do you think of them as yours?'

Lung Teep nodded.

'And Nai Tong, your land — do you think of it as yours?'

The second farmer nodded.

One by one, Inthorn turned to each of the farmers standing around him and asked them the same question. And one by one each answered in the same way.

'Well,' Inthorn said finally, 'they're not. That land out there belongs not to us but to some man who has never even held a crumb of soil between his fingers, yet who takes half of what we harvest, every year.' The farmer trickled another handful of grain through his fingers. Then he took a deep breath.

'No more,' Inthorn said. 'From this day on, I say that my land belongs to me. Yes, I will continue to give the landlord a part of my rice, because I recognise that some title deed somewhere has his name, and not mine, on it. But I will give him only what I say is fair, not what he demands. Young Ned here,' he nodded to the student, standing at the edge of the crowd, 'young Ned has been telling us of the many other villages where they pay only one-third of their rice crop as rent. He urges us to do the same . . .'

'I . . . I've always said that the decision was yours,' Ned said.

'Really? I thought, just a few days ago, you were telling me I had less spirit than a buffalo,' Inthorn said.

Ned looked acutely embarrassed.

Inthorn laughed. 'You were right, Ned. You said even a buffalo fights to have its own way sometimes. Well' I've been like a buffalo all these years, placidly following where the master led me.' He drew himself straighter, and his glance swept past the people around him. 'But this year I am starting to fight back. This year, I will resist the rent.'

A deep sigh, as strong as a gust of monsoon wind among the wild plumed grass, swept through the group. Jinda heard it, and felt a surge of pride in her father. So this is what makes him a leader, she thought. Not only has he just taken a major step for himself, but he has also convinced the others around him to take the same step for themselves.

In the twilight, a few skylarks winged towards the mountains, as tendrils of smoke from the first few fires in the village drifted up. Across the fields Jinda saw little Pinit coming towards them, tugging at his buffalo. With the hemp rope strung through its tender nostrils, it responded to every tug of Pinit's on the rope. How different from the spirited animal it had once been, Jinda thought. Pinit led the docile buffalo over to her father, who patted it affectionately.

She did not understand why, but watching them, Jinda was filled with a sudden sense of uneasiness.

Chapter 7

In the next few days, the villagers of Maekung talked of nothing else but the rent. Most of the farmers supported Inthorn, and had vowed to resist the rent too. But a few villagers held back, apprehensive and unsure that this step was the right one for them.

Suppose, they argued, that the landlord, or his agent Dusit, forced them off the land? Suppose the police came and arrested — even shot? — them all? Suppose Ned and those other students were really Communists, come to stir up trouble in their peaceful village?

It was no coincidence that the farmers who feared these possibilities most, were also the ones closest to Mau Chom. He was one of the few villagers in Maekung who owned his own land, and therefore never had to pay any rent. But this did not stop him from urging tenant farmers to continue paying the full rent.

To these farmers Inthorn spoke very patiently. 'One step at a time,' he would say. First they must feel that the land truly belongs to them, that they have a right to the rice they've grown on their land. 'Then,' he said, 'the next step is to hold back that part of the rice which we need for ourselves, for our children.' That's what the fight was about, after all, so that the little ones of Maekung need never go hungry again during the lean years.

During these few days, Jinda listened to her father talk. Out of a sense of loyalty to him, she would repeat his arguments when groups of young village girls and boys gathered to talk about the rent issue. But no matter how convincing she sounded, even to herself, she felt an underlying current of fear.

Her uneasiness deepened as Jinda began to realise how strongly her sister Dao — or rather, her father-in-law — opposed Inthorn's decision. Several times, Jinda would try to talk to Dao, but was rebuffed. Dao did not come home to visit anymore, and avoided her sister even when Jinda tried to seek her out at Mau Chom's house, or by the well, or during the lunch breaks they shared while harvesting in the fields. Jinda had seen her sister moody and sulking before, but Dao had never stayed aloof for so long.

What was also unusual was the absence of the rent-collector. Dusit must have known, through collecting rice from the villages nearby, and from reports filtered through to him from people like Mau Chom in this village, that the threshing in Maekung was complete. He should have been here by now, to haul off half their rice crop. Yet he had made no appearance.

For days, Jinda brooded about both her sister's aloofness, and Dusit's absence. Then she discovered that the two were not unrelated.

It happened one afternoon when she was walking by herself along the river, hoping to find some newly sprouted bamboo shoots. At a bend in the river almost two miles up from the village, while she was searching through the thick pile of fallen bamboo leaves, she saw a flash of red behind the bushes on the opposite bank.

A squirrel, Jinda thought at first. Then a figure moved into view, and Jinda stiffened. It was Dao, and she had on her best sarong, its white lotus flowers printed against a gay red. Instinctively, Jinda crouched down low, and watched.

Dao's hair was tied back, and held with a string of jasmine buds. Her face was clean and smooth, and she was even smiling a little.

How pretty she is, Jinda thought with pleasant surprise. She looks as lovely as the mornings when Ghan was courting her.

Curiously, Jinda watched her sister walk along the opposite bank. It was much more overgrown there, and Dao had to move cautiously to avoid the thorns on the tropical mimosa bushes. Once, Jinda noted, her sister even looked over her shoulder furtively, as if afraid she was being watched.

Near a large willow tree with trailing branches. Dao paused and called out softly. The overhanging branches parted, and a man stepped out. Even from a distance, Jinda could see that he was tall and broad-shouldered, solidly built. He was also dressed stylishly, in city clothes and shiny leather city shoes. But his face was in shadow, and Jinda could not see who he was.

Languidly, the man stretched out a hand and tried to pull Dao to him. She pulled her hand away, and slapped him lightly. Jinda heard her sister's high, light laugh across the river.

She's flirting, Jinda thought incredulously. This wasn't the sulky, withdrawn young woman she'd seen these last few weeks.

Dao laughed again, a musical trilling sound that echoed across the river. Jinda was mystified. Why? she wondered. Who was Dao with? Had her husband Ghan come back to her after all?

The man half-hidden behind the willow branches reached out for Dao again, and this time she did not resist. Hand in hand, they walked into the wall of leaves. The man was laughing, his teeth flashing in the dappled sunlight. Jinda caught a glimpse of his face.

No, he was not Ghan.

Dazed, Jinda stared after him, but the trailing leaves hid the two of them from view. Behind the leafy canopy, the man's shiny leather shoes edged closer to Dao's bare feet, and stayed there. They were standing very close together, Jinda thought.

There were quiet murmurs, at first deep and serious, but then giving way to Dao's breathless giggles. Jinda recognised her sister's laughing protests: she was enjoying herself.

For a long time Jinda watched them, or watched what she could see of them. Once Dao rubbed her foot against the man's sock, and pushed it down. Then the leather shoe twined around Dao's leg, and stayed there.

In the stillness she could hear nothing except the dreary hum of the cicadas, and the occasional wail of a wild gibbon.

Finally the delicate branches parted, and the man stepped through. For the brief instance that Jinda saw him, she thought he looked familiar, but she wasn't sure who he was. Then, without once looking behind him, the man walked up the river bank, and onto the hillside behind it.

Only then did Dao emerge from behind the same branches. She looked flushed and dishevelled. Her hair was untied, and the string of jasmine buds was in disarray.

Jinda stepped out of the shelter of her bamboo grove, and waved.

'Jinda! What're you doing here?' Dao had been trying to retie the jasmine garland around her hair, and her arms were lifted high above her head.

Watching you, sister, Jinda wanted to shout. But she only smiled weakly and said, 'Collecting bamboo shoots.'

'How . . . how long have you been here?' Dao asked nervously.

'Not long at all. I just got here.'

'You're lying!' Dao cried. 'You saw us!'

'No, I didn't,' Jinda said, then realised too late how guilty her denial sounded.

Stumbling over to grasp Jinda's arm, Dao said, 'It isn't what you think. We were just talking business. And besides, he means well. Dusit really means well.'

Dusit, Jinda thought. Of course, that's why he looked familiar. As the rent-collector, Jinda saw him only once a year, but his pale, glossy face was distinctive.

'What business could you possibly have with Dusit?' Jinda asked now.

'He's offering to pay Father five thousand baht, if Father will just agree to pay the full rent as usual. Just think, Jinda, five thousand baht!'

'Father will never be bribed,' Jinda said hotly. 'Who does Dusit think he's dealing with? The oily fat crook!'

Dao flinched, as if she'd been slapped. 'Mr. Dusit's not a crook,' she said, 'and he's not fat!'

Jinda stared at her sister. 'So that's it,' she said slowly. 'You and that Dusit. This isn't the first time, is it, that you've crept out to meet him? All this time you were avoiding me, you were actually just seeing him, weren't you? Oh Dao, how could you?'

'Mau Chom made me,' Dao said petulantly. 'When he saw how you were all falling under the spell of those Communist students, he made me go and tell Mr. Dusit.'

'And then?'

Dao tucked a stray wisp of hair behind her hair. 'And then nothing, sister,' she said. 'I told Mr. Dusit and he offered to help Father out this harvest, that's all.'

'That's not all,' Jinda said grimly. 'How many times have you met Mr. Dusit since then?'

'That's none of your business.'

'But why are you so friendly with him. Dao? You know he's the rent collector; you know how everyone else in Maekung feels about him. He's not one of us. He belongs on their side.'

'He's a very friendly, kind man,' Dao said defensively.

A jasmine bud was dangling precariously from a loose

strand of Dao's hair. Jinda plucked it out and held it in her hand. 'How kind can he be, sister?' she asked.

Dao snatched the jasmine from Jinda's hand. 'Much kinder than you are, anyway!' she shouted, and walked off, on the trail heading back towards the village.

Three days later, Dusit made his appearance in the village. His arrival was announced by two children who came racing down the path one morning, screaming at the top of their lungs, that the truck was coming.

Within minutes, dozens of villagers had lined the street leading from the Outbound Path to the main threshing ground. Jinda was among them.

Soon a faint roar could be heard, and around the mountainside, a blue truck suddenly careered into view. Swerving so sharply that a few children had to jump into the ditch, the truck sped down the narrow road and pulled to a grinding stop by the threshing ground.

Dusit opened the door and swung himself from the running board to the ground. In his leather boots, he looked tall and commanding as he leaned on the hood of his truck. His clothes were bright and new, and if his pants seemed tight around the waist, that only added to his general sense of well-being. He seemed almost to glisten in the sun, Jinda thought, as his shiny belt buckle, his watch and a large ring all caught the sunlight and gleamed. Yes, Jinda thought, I can see why Dao would find him attractive.

Dusit propped his sunglasses rakishly over his forehead, and surveyed the villagers. His eyes came to rest on one face: Dao's. He smiled boldly at her.

Dao flushed, aware of the villagers' curious gaze. Then deliberately, almost defiantly, she smiled back at Dusit.

'Well, well, well,' he said genially, turning now to the rest of the villagers. 'Here I am again. As regular as the bullfrogs after the rains.' He laughed again, and a gold tooth sparkled in the sun.

66

No one spoke.

Pushing aside a few farmers, he made his way to the threshing ground. There, he circled the pile of rice slowly. 'Pretty poor harvest,' he said to Inthorn genially. 'Wouldn't have stashed any of it away in your rice barns already, would you?' He laughed loudly, as if in appreciation of his own joke.

No one laughed with him.

Dusit shrugged. 'All right. Let's see, there's no fertilizer or ploughing costs advanced in this village, so it's half and half of the crop, as usual. Come on, let's get moving. We haven't got all day.' He nodded to the assistant he had brought with him, and the man obligingly propped a board up onto the back of the truck, making a gangplank between the ground and the truck.

Silently, Nai Tong lifted a bushel basket of rice from the threshing ground, and walked up the plank, then poured the rice onto the floor of the empty truck. The sound of the rice hitting the metal was like a sudden gust of rain. Then, Lung Wan hoisted another basket of rice, and headed silently across the road to the village granary, a wattle bin built on stilts high off the ground.

Inthorn was next. He lifted a third basket, balancing it between his strong neck and hands, and started walking. Slowly and deliberately, he passed the blue truck and crossed the road to the granary.

'Hey! You there! Come back with that! It's the third basket!' Dusit shouted after him.

Inthorn kept walking.

'Somebody stop that fool! He'll get the order all confused. That's my basket!'

No one moved. In complete silence they stood watching Inthorn pour the rice into the village bin.

When Inthorn came back to the threshing ground, Dusit knocked the empty basket out of his hands. 'What's the matter with you?' he demanded. 'You know damn well the first basket's mine, the second's yours, and the third one's for me!'

'Not anymore,' Inthorn answered. 'The rent's too high. From now on, it's one basket for you, and two for us.'

'Who says?'

'I do,' Inthorn said quietly.

'You're crazy! Who're you to tell me what the rent is? You're just a farmer!'

'Yes, I am a farmer,' Inthorn said. 'I plough the land. I sow the seed rice. I transplant the seedlings. I water and weed the fields, I harvest and thresh and winnow.' He drew himself up very straight and looked down at the rent-collector. 'Yes, I farm the land. What do you do?' he asked.

'I . . . I . . . you . . .' Dusit was speechless. Bright patches of purple mottled his face. 'You can't do this!' he spluttered.

'Watch us,' Inthorn said. He nodded at three farmers. Immediately they filled up their baskets. One walked towards the truck, the other two headed for the rice bin.

The division of the rice went on quickly and smoothly after that. For the first time, it seemed to Jinda that the farmers carried out the task with a song in their hearts. 'Watch us,' Inthorn had said. Watch us, watch us, watch us: the bare feet slapped lightly on the path, bringing basket after basket of rice into the village granary.

Dusit stood and watched. He was sweating heavily, and his shirt was already soaked through. His face, too, glistened with sweat, and looked pale and oily. Like the skin of boiled chickens hung from noodle stalls, Jinda thought, watching him.

At last, he retreated to the shade of a bamboo grove where Dao was standing. The other village girls in the shade quickly scattered as Dusit approached, but Dao remained.

Although Dusit was careful to keep his voice low, Jinda moved close enough to catch a few phrases.

'What happened? You said you could persuade him,' the rent-collector said.

Dao's voice was inaudible, but she talked to him in quick, urgent whispers, her head bent as if she was ashamed.

Dusit edged even closer to Dao, and now reached out to brush aside a strand of hair from her neck. It was a slow, intimate gesture. Conscious of the village girls watching them, Dao moved away, but Dusit grabbed her wrist.

Laughing nervously, Dao tried to shake him off. He only pulled her closer.

As they were grappling, Inthorn passed by, carrying a basket of rice on his shoulder. He noticed the commotion, and paused to see what was going on. When he saw Dusit holding onto Dao, he dropped the basket and ran over to them.

'What do you think you're doing?' he shouted. 'Keep your hands off my daughter!'

Startled, Dusit dropped Dao's hand.

Inthorn shouldered Dao aside, and confronted the rent collector. 'You've come to collect our rice,' he said, trying to control his anger. 'Not to molest our women!'

'Father,' Dao began, 'Mr. Dusit was just . . .'

'Enough out of you,' Inthorn said fiercely turning to Dao. 'Go home. Right now. And stay there!'

Her eyes bright with tears of humiliation, Dao backed away.

Inthorn lifted the basket of rice, 'I don't ever want to see you with her again.' Hoisting the basket on his shoulders, he walked away, a powerful spring to his step.

Jinda watched Dusit's face as he looked after the farmer. It was contorted with rage, and the expression of naked hatred on it frightened Jinda.

By late afternoon, the farmers had finished and the threshing ground was bare. The blue truck still had more space for rice, but the village granary was fuller than it had been for many years.

Just before leaving, Dusit stalked over to Inthorn and muttered something under his breath. The farmer looked troubled, but when Jinda asked him what Dusit had said, her father only smiled vaguely, and told her not to worry.

Then, as everyone watched, Dusit and his assistant climbed back into their truck. With a grinding of gears, they drove off in a cloud of dust.

'We did it!' Inthorn said, as he watched the truck disappear. He looked at the triumphant faces of the other farmers around him, and broke out into a broad smile himself. 'All right, let's celebrate. We'll have a village feast tonight. Come on!'

And so that night, the whole village celebrated.

Girls threaded jasmine buds in their hair, and even Jinda's grandmother playfully tucked a bright red hibiscus in her bun. Fresh offerings of rice and flowers were placed on the altars of the spirit houses, and a new sash of homespun cotton was tied around the trunk of the old raintree sacred to the village. A bonfire was built near the raintree, and a few of the less scrawny chickens were killed to make a big pot of chicken curry.

Nai Tong and his brother brought out their long drum, and Lung Kam played his three-stringed lute. Late into the night, the villagers danced to this simple music, weaving upturned palms in the traditional Ramwong dance.

Jinda danced with Vichien and other village boys, but out of the corner of her eyes she kept watching Ned.

He had become the hero of the village. Continually surrounded by giggling village girls, or pulled into quiet corners to confer with the men, he seemed to be the centre of attention of every group he was with.

Yet he managed to extricate himself from the crowd, and find time for one dance around the fire. And this one dance was with Jinda. He sought her out among the girls she was standing with, and said smilingly. 'I'm not very good at dancing, but maybe you could teach me.'

Eyes glowing, trying very hard to keep her smile demure,

Jinda danced around the fire with him. Their bare arms, tracing sinuous patterns in the air, gleamed in the firelight. And although they never once touched, Jinda was acutely aware of his nearness.

'You're very beautiful tonight, Jinda,' he murmured, as they circled each other.

Jinda's cheeks burned. The fire suddenly seemed too bright. she danced away from it, and Ned followed. As they weaved their way past other dancers, their fingertips touched. Ned held on briefly to Jinda's hand. She felt strangely excited and calm at the same time.

The firelight stroked his face with light and shadows. And she wished her fingertips were like the firelight, long and light and playful on his cheeks and throat.

Finally the drumming slowed down, and the music from the lute trailed away. Like the other couples, Jinda and Ned stopped dancing and separated, but not before Ned's hand brushed against hers one last time.

As Jinda left the bonfire, she felt like laughing out loud. The cool night air caressed her face, and she felt happier than she had ever been.

Then she saw Dao. Under the raintree, she was standing very straight, very slim, and very alone. A single jasmine blossom in her hair, she looked fragile and lovely. Jinda realised then that she hadn't seen Dao dance once all evening. The villagers were all avoiding her, as if she belonged to Dusit already, Jinda thought. She waved to her sister, but Dao turned abruptly and walked away into the night.

A gust of wind blew past. The night had turned chilly and Jinda shivered.

The mood of celebration did not last long in Maekung. Within two days Dusit was back again.

The time he parked his truck right next to Inthorn's house, and in the seat next to him was a stranger wearing a khaki uniform: a police officer. Jinda watched from the gate as the two men got out.

'Is this the village headman's house?' the policeman asked.

'I am the headman,' Inthorn said. He had been sawing a teak log to replace some rotten planks on the verandah, and he carried his saw with him as he stepped foreward.

Dusit eyed the saw nervously. 'Put that thing down,' he said.

The farmer smiled slightly as he did so.

'So you're the man who plotted to resist paying the rent?' the policeman asked him.

'There was no plot,' Inthorn replied calmly. 'Every man made the decision for himself.' Nai Tong and several farmers had gathered by the truck to listen, and a few of them nodded in agreement at Inthorn's words.

'That's beside the point,' Dusit said. 'As the headman, you're the one responsible. You refused to pay the full rent, and so,' Dusit paused and smiled grimly, 'I'm having you arrested.' He nodded at the police officer, who stepped towards Inthorn.

'Wait!' Ned pushed his way through the crowd and stood between Inthorn and the policeman. 'You can't arrest him,' he said. 'Headman Inthorn hasn't broken any law.'

'The hell he hasn't! He didn't pay his rent!'

'He paid one-third of his harvest yield. That's not breaking the rent law.'

'What do you know about the rent law?' Dusit demanded. 'I've been collecting rent since before you were weaned, you pup! I know when someone hasn't paid his legal rent.'

'No, you don't,' Ned said evenly. And in his calm, clear voice, he told Dusit what he had told the villagers many times before, that the new government was considering passing a new land-rent law, limiting the rent to one-third of the crop. 'Until this proposal is either passed or rejected,'

Ned said, there is no legal rent. And therefore headman Inthorn has broken no law.'

Dusit listened open-mouthed. Beads of sweat were glistening on his forehead, and he flicked them off. 'How do you know all this?' he asked.

'Does that matter?' Ned countered. 'It's true, isn't it?' He looked at both Dusit and the policeman. 'The point is that no law stipulates that a tenant farmer must pay one-half of his crop as rent.'

The policeman frowned. 'Who are you?' he asked abruptly. 'You don't talk like a peasant. You wouldn't happen to be some Communist come sneaking over from Vietnam or Laos, would you?'

'I am as Thai as you are,' Ned said.

'Commies consider themselves Thai,' the policeman retorted. 'Even though they've been trained in Vietnam.'

'I am not a Communist,' Ned said.

'Yeh?' Dusit eyed him sceptically. 'So who are you?'

For a second Ned hesitated. 'I'm a student,' he finally said, 'from Thammasart University.'

The policeman and Dusit exchanged a quick look. 'I thought so,' the officer said. And although he sounded hostile, his voice held a grudging respect. After all, hadn't Thammasart students organized hundreds of thousands of people in Bangkok to topple the military government? Students were a powerful new force in the country.

The police officer drew Dusit aside now. 'I can't arrest your Inthorn now,' he said quietly. 'Not with a Thammasart student watching. He's right: refusing to pay the traditional rent isn't a good enough reason to arrest your headman.'

'Then arrest him for some reason!' Dusit snapped. 'These peasants are always breaking some law or other.'

Dusit suddenly smiled, and sauntered over to the log that Inthorn had been sawing. 'Looks like a good solid log you've got here,' he said with studied casualness. 'Teak, right?'

Inthorn nodded warily.

'Cut it yourself?'

Again the farmer nodded.

'Where'd you cut it from?'

Inthorn hesitated, looking to Ned uncertainly. But the student looked puzzled too.

'You got it from the Chiang Dao mountains, didn't you?' Dusit asked pleasantly. 'Where most of the teak trees have been cut already?'

'Wait a minute,' Inthorn protested. 'I only cut down one or two logs. Those mountains were stripped bare by the timber company.'

'Oh, I know that,' Dusit said. 'The timber company has a concession to cut those trees. You don't.' His voice suddenly became harsh. 'You broke the law cutting that teak tree down.' He nodded at the police officer. 'All right, arrest him, for being in possession of illegal timber.'

There was a stunned silence. It had all happened so fast that no one could do anything.

Dusit laughed at Ned. 'Want to consult your Thammasart textbooks about that, son?' he said. 'I haven't been to your fancy university, but I know one thing: illegal timbering can get a man up to fifteen years in prison.'

'Fifteen years!' The shocked whisper reverberated through the crowd. 'No! Not Inthorn!'

Inthorn looked around him desperately. The policeman stepped forward and grasped his arm.

'No!' Inthorn cried, wrenching free. The next moment he had turned and dashed through the crowd.

The villagers parted quickly to let him through. Jinda saw him plunge out of the crowd and make for the fields beyond.

Run, Father, run, she urged him silently. Faster!

And then a shot rang out.

The policeman had dashed after Inthorn, and was waving a pistol in the air. 'Stop or I'll shoot,' he cried. 'Stop!'

But the farmer only ran faster. Tall and lean, Inthorn sprinted towards the stubbled fields. He was heading for the river, Jinda realised. Once there, in the thick underbush, he would be safe. Run, she urged him in her heart. Run, run!

There was a second shot.

Then utter silence.

Inthorn had crumpled up in a far corner of the rice field. Jinda saw him sink slowly to his knees, a small black dot among the brown. For what seemed an endless moment, he did not move. Then, finally, Jinda saw her father very slowly straighten up.

'Thanks be to Buddha,' Jinda breathed. And the soft prayer of relief echoed through the crowd. 'Thanks, thanks be to the Buddha.'

The police officer ran over to Inthorn, and took hold of the farmer's elbow to steady him. Inthorn shook him off. Alone, head held high, he walked unsteadily back towards the crowd.

Looking dazed, he allowed the policeman to steer him through the stunned crowd and into the truck.

When he was seated, Jinda walked over and gently examined his hand. The wound did not look serious. The bullet had just grazed the back of his hand, and the stitches on his palm were still intact. Carefully, Jinda poured some water over the wound, rinsing off the dirt and blood. Pinit came over with another ladleful of water, and she used that too. When it was reasonably clean, she looked around for something to bandage it with.

Pinit understood. Quickly he stripped off his T-shirt, and handed it to his sister. She wrapped their father's hand in it, and knotted the sleeves together. It made a loose, clumsy bandage, but at least it would keep the dust off during the long ride ahead.

Chapter 8

Almost a month had passed since Inthorn's arrest, and still no one had been able to visit him in prison. Jinda had tried several times to see him, but without success. Together with Ned, she had made the long trip into town, and tried to petition for permission to see Inthorn. But each time they were passed on from one bureaucrat to another, and it was only because Ned was obviously educated and from Bangkok that they got to see anyone at all. Medical treatment? Sorry, but a doctor was called in only in exceptional cases. Visits? Sorry, but orders from higher up stipulated that Inthorn Boonrueng's case was special, and he could receive no visit from anyone. Bail? Sorry, he didn't qualify for it. An early trial? Sorry, but there was a back-log of cases awaiting trial, and he would have to wait his turn.

The waiting seemed interminable to Jinda. It was almost three weeks later, that she finally received a letter from the prison authorities informing her that she could visit her father.

She held the precious letter in her hands now, as she sat in the oxcart. She was finally on her way to visit her father and, under the letter, Jinda held a cloth bundle on her lap. Inside were home-made coconut patties, and a package of capsules given her by Sri, medicine to prevent Inthorn's wound from festering. Jinda hugged both the letter and the cloth bundle to her.

In the morning light, the paved road stretched out ahead of the oxcart, shiny as new lacquer. Patches of dew glistened like shattered glass on the weeds alongside the path.

The team of oxen ambled on, their wooden bells clanking melodically. It was still chilly; Jinda edged her toes over to a patch of sunlight on the wooden cart, and wiggled them. Dao slid her foot over and gently nudged Jinda's toes with her own. Jinda looked up and smiled at her. If nothing else, she thought, at least Father's arrest had brought her closer to her sister.

The two of them were sitting next to each other, their backs resting against the sides of the cart. With each jolt from a rut, their shoulders would bump together. Each time they touched, Jinda could feel the warmth of Dao's breath on her neck, and this reminded her of the early mornings when she and her sister would lie, curled sleepily against each other, in their mosquito net.

At the back of the cart, sitting close together for warmth, were Ned and two farmers, Lung Teep and Nai Tong. They had insisted on coming, although they weren't sure whether or not they would be allowed into the prison to see Inthorn.

The sun was high in the sky by the time their oxcart entered the outskirts of the town where the prison was. At first sight, the atmosphere inside seemed more like that of a carnival than a prison. Makeshift stalls set under bright sun-umbrellas dotted the square, and people milled about, clutching children and shopping baskets behind them. It looked much like any marketplace — except for the tall stone walls looming behind the gay umbrellas.

Nai Tong parked the oxcart under a tree, and Jinda and the others all got off and walked to the market.

Under their striped canopies, wooden stalls displayed piles of fried chicken and sticky rice. Jinda paused by a mound of tangerines and eyed them wistfully. Her father was very fond of tangerines. If only she could afford to buy him a bag! But Ned came up behind her and steered her past them to a booth right in front of the prison gates.

This was where visitors registered their names. Carefully, Jinda wrote down her own name on a ledger. The officer who manned the booth looked at her with interest.

'Jinda Boonrueng,' he read aloud, 'Are you Inthorn's daughter?'

Jinda nodded.

'He's a special case. Allowed only one visit, and needs special permission even for that once. Have you got permission?'

Jinda produced the letter and held her breath.

The man in the booth nodded. 'All right,' he said, 'Only this once, and only his immediate relatives.'

Ned tried to argue with him, but in vain. Neither he nor any of the farmers would be allowed in to see Inthorn.

So Dao signed her name under Jinda's, and they all went back to the market to wait.

Jinda was nervous, but Dao seemed even more edgy. After a few minutes, she excused herself from the group, saying that she had some business to attend to. When Jinda offered to accompany her, to her surprise, Dao firmly insisted on going alone.

Left to wander around without her sister, Jinda listened idly to the snatches of conversation in the market around her. A few of the people milling there were farmers, but most of them were obviously from the city. A plump woman with long, blood-red fingernails was complaining that the police had closed her husband's gambling den again, and two pock-marked teenagers were comparing the price of heroin. Were these the kind of people that her father was thrown in with, Jinda wondered?

'Hear it's getting more and more crowded in there,' one of the teenagers said.

'That's hard to believe. Packed twenty-eight to a cell when I was in there. Couldn't turn over at night without waking the guy next to you.'

'Well, at least you were never in the Blackhouse,' the first man said. 'I was shut up there for a week once. No windows, no light, no toilet. Pure hell, that's what it was. When I came out the sunlight was so bright I couldn't see properly for hours.'

'Well, I saw you,' the other man laughed. 'Your skin covered with sores, blisters peeling off — God, you looked awful!'

Despite the noon sun, Jinda shuddered. What would her father be like? How had he been treated these last four weeks?

There was a loud crackling noise, and the loudspeaker strung up above the market square, started blaring. The crowd quieted down as a metallic voice boomed out.

'Group Twelve,' it announced, 'Visitors of the following inmates can now proceed inside the prison. Single file at the gate, please.'

A list of names was read out. When the voice announced 'Inthorn Boonrueng', Jinda looked around for Dao, but her sister was nowhere in sight.

It was only when most of the other visitors had already filed inside the prison gates, that Dao rejoined her. She looked tired, and was out of breath.

'Where've you been?' Jinda asked impatiently.

'Just to the Prasingh temple,' Dao said. 'To light some incense and say a prayer for Father.'

'That's all?'

'Yes, and stop staring at me,' Dao frowned. But Jinda could not take her eyes from Dao's hand. On the middle finger was a new ring, a thick gold band set with a red stone. Jinda tried to think where she had seen it before, and then suddenly she remembered. Dusit had been wearing that same ring on his plump little finger.

Jinda turned away, and did not ask Dao any more questions.

There was a narrow doorway in the prison wall, and they were ushered through it by uniformed guards. Jinda and Dao held hands as they stumbled along with the crowd through a dim corridor to an equally dim, but much bigger room.

The room was partitioned into three sections by iron bars, the middle section empty except for two prison guards. Sandwiching the prison guards were the visitors on one side of the iron bars, and the prisoners on the other side. There was about six feet between the two groups.

Jinda had only a second to take this in. She was shoved along by the crowd towards the iron grill, fighting blindly for a space next to the bars. A little boy wriggled past her and prised himself between her and Dao, clinging tightly to the bars.

At first Jinda could not see her father. Separated by the gap, the men on the far side of the partition looked dark and indistinct.

Then she heard Dao's shrill cry. 'Father! Over here! Here!'

Desperately, Jinda scanned the faces of the caged men, and at last found her father.

He had aged. In the dim light his face looked haggard and sallow, and long lines tugged down the corners of his mouth. His clothes hung loosely on him, and one bare shoulder, thin and smooth, protruded from his torn shirt. His trousers too, looked baggier, hanging loosely from his waist.

And around his ankles were iron shackles.

Jinda stared at the shackles in horror. Thick iron bands were clamped around each ankle, chaining them together. He was the only prisoner to have shackles. Could he be considered more of a menace than heroin pushers and gamblers?

He snuffled over towards them now, the shackles clanking against the metal bars as he moved. In his hands he held part of the chain, lifting it as he dragged each foot along. Carefully he edged his way over until he was directly in front of them. Jinda saw that he was trying to say something to them, but the noise of the others shouting drowned him out.

Clutching the metal bars, he mouthed the same words over and over. Only by watching his lips move, could Jinda make out what he was saying. 'Get me out,' he said. 'Get me out!'

'We're trying,' Dao shouted above the tumult of other voices. 'Maybe Dusit can help . . .'

Inthorn shook his head vehemently. 'Not Dusit!' he shouted. 'What about Ned?'

'He's trying!' Jinda shouted back. 'But the bureaucracy...'

Again Inthorn shook his head. 'Too long,' he called out. 'My hand . . . it's getting bad.' He looked extremely tired. 'Please hurry,' he said.

After, a bell was shrilling. Guards prised visitors away from the bars, but many clung on, screaming last messages at the men on the other side. A few thrust bags of food and cigarettes at the guards, who grudgingly ferried them across to the prisoners. Jinda too, thrust her parcel of food and medicine at a guard, and begged him to pass it on to her father.

'What's in it?' the guard demanded, taking the parcel.

'Coconut patties,' Jinda said, then, seeing that he was unwrapping it anyway, added, 'and some medicine.'

The guard pulled out the package of blue and white pills. 'Sorry, drugs are contraband,' he said.

'But it's for my father's hand!' Jinda said loudly. 'He's hurt!'

The guard shrugged. 'Rules are rules,' he said, pocketing the pills. He sauntered across to the other partition, and handed the parcel over to Inthorn.

'Father!' Jinda called to him.

He took the parcel, and looked up at Jinda. For the first time during the visit, he smiled. 'Jinda,' he called, stretching out his arms to her, past the bars. Only then did Jinda see that his left hand was swollen and bloodstained. The T-shirt bandage she had wrapped around his wrist was in shreds, and stained with pus and dirt.

'Father, your medicine!' Jinda shouted.

'Medicine?' the farmer echoed faintly, looking relieved. 'Good, thank Sri . . .'

'No, Father, they took . . .'

But a guard was pulling her father away, dragging him by the chain through a narrow doorway on the other side. Jinda watched him, teeth clenched to stop the tears. She leaned her head against the bars, and the metal was cold and smooth against her forehead. Oh father, she thought, your hand — what will happen to your poor hand?

Someone gently pulled her away from the bars. She found herself held for a brief moment in Dao's arms. Choking back her sobs, she clung to her sister, and together they made their way down the dark corridor towards the square of sunlight at the far end.

The sunlight seemed glaring after they emerged from the prison gates. The men were waiting for them, but Jinda was too shaken at first to say very much. She had known that her father would suffer in prison, but she had not expected him to have aged so drastically.

Ned put his hand gently on her arm. 'Don't be discouraged,' he said. 'Your father isn't alone. Look what I found.' He handed her a magazine, and she glanced at it. 'It came out just last week.'

On the cover of the magazine was a photograph of some ordinary farmers threshing rice, but emblazoned across it were the words: 'Inthorn Boonrueng, latest hero in the rent resistance movement.'

'It's a new magazine,' Ned was telling her. 'Very

progressive. Jongrak and Pat know the editor. They sent in the material about your father.'

Jinda leafed through the article. Quickly she skimmed through the long phrases — 'symbol of the oppressed peasants,' 'revolutionary leader of the farmer's movement,' 'challenger of the exploitative and feudalistic land rent system'. Was this her father they were writing about?

What about his leg irons, Jinda thought. Why don't they write of those leg irons and how they've scraped his ankles raw? She felt a surge of resentment against Ned. You got Father in, she wanted to shout at him. Why can't you get him out? Is this all you're capable of doing, writing fancy long words about him?

Wordlessly, she handed the magazine back and walked away.

For the next week, Jinda kept her distance from the students. She knew that this would be their last few days in Maekung, since the universities in Bangkok were about to start their new term. Holidays over, Jinda thought bitterly, time to go back to school now.

Ever since she had seen that magazine article about her father, Jinda had felt a deep hostility towards the students, thinking that she and her family had been deliberately used by them. She resented Jongrak and Pat, for sending that information about her father to some Bangkok news office. She resented the shy, quiet Sri, who regarded politics as a way to change the public health system. And most of all, she resented Ned.

During all the weeks since her father's arrest, Ned had pretended to like her and help care for her. Ned had made the rounds of the bureaucracy with her, trying to get permission to see her father. Ned had waited for her after Sri's clinic, and walked her home, talking to her of his day and asking about hers. Ned had helped her fetch water from the well, and empty her buckets into the earthenware jars on the verandah. Ned had helped Pinit built the weed-fire next to the buffalo, to keep the mosquitoes at bay. Ned had, Jinda thought, become a friend.

But he obviously did not see her in the same light. She was not Jinda, but some political symbol he wanted to make use of for his own political purposes. She was not a young woman, but an 'oppressed peasant', just as her father was not Inthorn, but had become by some political sleight of hand, 'a symbol of oppressed farmers'.

Jinda knew, the day before the students were to leave, that Ned had been looking for her. But she had managed to dodge him, and he had been so taken up with other farmers who wanted to confer with him, that the two of them never had a moment alone.

Still, that last evening, Ned stood outside her gate and called softly to her. Jinda was sweeping the verandah just then, and a few specks of dust from the cracks of the floorboards rained down on him.

'Can you come down for a walk?' he asked Jinda.

Jinda swept furiously, sending a storm of dust on his head. 'Sorry, I'm busy,' she said.

'Wait, who's that calling you?' Jinda's grandmother shouted, peering out of the kitchen.

'Nobody,' Jinda said.

'It's Ned, and I want to talk to your granddaughter,' Ned shouted. He looked very much the way he had that first evening at the river bank, with his shirt unbuttoned, and his chest muscled and brown.

The old woman stepped out onto the verandah, and motioned for him to come up. 'Ned,' she said. 'Don't just stand there. Come up. There's something I want to say to you.'

Ned climbed up the wooden steps to the verandah, and sat down in front of Jinda's grandmother. Jinda herself was careful to keep in the background.

The old woman motioned for Ned to sit closer to her. 'I hear you're leaving tomorrow,' she said.

Ned nodded.

'When are you coming back?'

Jinda's pulse quickened. How often she had wanted to ask that question herself, but never dared. For once, she was glad that her grandmother was so blunt.

'I don't know,' Ned said. He turned and looked at Jinda. 'But I hope it will be soon.'

'Well, I hope so too. You're a good boy, even if you do have some strange ideas.' The old woman was silent for a long moment. 'You've managed to change more things in this village in two months, than anybody has done for twenty years. Let's just hope these changes will turn out well.' Then, carefully, she picked up a thick strand of homespun cotton, and held out her hand to Ned.

'Now give me your hand,' Jinda's grandmother told him. 'The left one.'

Ned obediently stretched out his left hand towards the old woman.

Rocking gently from side to side, she murmured a long blessing, her voice undulating with the same rhythm as her rocking. It was the traditional leave-taking ceremony of the North, and the old woman was calling back all the wayward spirits of Ned's that might have strayed into the village, so that he could leave a whole man. Spirit of Ned's eyes, that have seen good things and bad in Maekung, come back to him, she chanted. Spirit of Ned's ears, she continued, winding the white thread round and round his wrist.

Solemnly her grandmother tied a knot on the thread bracelet. The strands of cotton looked very white against his brown arm. 'Bless you, child,' she said. 'And come back soon.' She got up then, and walked back into the kitchen.

Ned stroked his thread bracelet, and looked across the verandah at Jinda. 'I meant it, you know,' he said quietly. 'I will be back soon. One of the main reasons I'm going back to Bangkok at all is to try and get Inthorn released from prison. Jinda, look at me. Don't you believe me?'

Jinda kept looking at her hands. 'What does going to Bangkok have to do with my father?' she said.

'You've seen how useless it's been, trying to get some bureaucrat in town to help. This whole rent issue has become too controversial, too widespread for the provincial officers to handle. We've got to raise the issue with the politicians in Bangkok.'

Politics again, Jinda thought bitterly: games of the city.

'I know what you're thinking,' Ned said quietly. 'And I think I understand, but I want to tell you that you're wrong. Come on, Jinda, let's go for a walk. Give me a chance to explain.' He looked at her earnestly. 'Please, Jinda, it means a lot to me.'

In silence, they walked down the small trail behind the village, and up the hillside. Ned did not talk much, perhaps sensing that Jinda mistrusted his fluency. Instead, he pointed to the many small things in Maekung he would miss, and tried to tell her how happy he had been in the village. 'It's been a wonderful two months, Jinda,' he said, 'I won't pretend there hasn't been unhappiness too, but I've been happier here . . . with you, than I can remember ever being. I feel at home here.'

And he told her too, of the drab, cluttered rooms he lived in, in Bangkok, and of long days spent in meetings and classes. 'Sometimes I feel so frustrated, Jinda, as if nothing around me is real. And I'm a bit scared that after I've been back there too long, Maekung, and even you, will seem unreal.'

Jinda said nothing, but she understood, because she had faced the same fear in herself as well: that after Ned left, she would feel as if he had never really come and shared her life these few short months.

'I know you can't promise anything now, Jinda,' he said, 'but I'd like you to consider coming to Bangkok soon. I plan to organise a rally around the issue of the land rent in a few months, and I'd like you to come then, maybe give a speech.'

Jinda stiffened. 'There are many others who can give speeches better than I could,' she said. 'You don't need me in Bangkok for that.'

They had made a full circle, and come back to Jinda's house, where they were standing under the mango tree, beside the little spirit house. Ned bent down and picked a single amaranth blossom from a nearby bush, and handed it to Jinda. 'You're right, Jinda. Thank you for making me say this: I'm asking you to come to Bangkok because I want to see you again, and soon.'

Jinda felt a sudden lift of her spirits. She looked up at him and smiled. 'Now that,' she said, accepting the small purple flower from him, 'is a much better reason.'

He cupped her chin lightly in his hand, and tilted her face up to him. 'It'd make it much easier for me to leave,' he said, 'if you'd promise me you'd come to Bangkok soon.'

Jinda laughed, even though her throat felt dry. 'I don't want,' she said softly, 'to make it easy for you to leave.'

For a brief moment he put his arms around her, and held her against him. Then, almost abruptly, he let her go. 'I'll see you in Bangkok,' he said, and walked away quickly.

Jinda watched Ned pause by the little spirit house, and quickly, bow his head towards it, just once. Then he was out of the gate and walking down the dusty little trail, his white shirt billowing in the evening breeze.

The air was cool and fresh, and the sun was just setting. A single shaft of light illuminated the little spirit house. It looked unkempt and forlorn. Jinda went over to it, and hesitated.

Then she stood on tiptoe, and dusted its altar with a square of banana leaf. She placed the amaranth blossom Ned had just given her on the altar. There was no incense or fruit, no other offering there. The roof of the spirit house sagged, and its wooden walls were warped.

She remembered the day her father had built it, sawing the little pieces of wood and carefully nailing them together.

Her mother, her grandmother, her sister, and Jinda herself had all stood around, admiring his workmanship. When he finished, they had all helped to put in offerings until it was brimming over with fruits and flowers. How full and happy it had looked then!

And how empty, how terribly empty, the little house seemed now. Slowly Jinda turned and walked away from it, without bowing.

Chapter 9

The train started with a small jerk. Up ahead, the whistle blew piercingly as the engines slowly picked up speed. On the platform outside, boys balancing trays of boiled peanuts and coconut cakes trotted alongside the train, trying to hawk one last sale. Standing a little away from them, knots of people waved goodbye.

Jinda settled back into her seat. She had no one sending her off, and no money to buy any snacks with. She watched the vendors, the platform and finally the train station recede into the distance as the wheels churned faster and faster underneath her.

Soon they were out in the country. The parched land stretched away in both directions, burnished a flat gold by the late afternoon sun. It was April, the heart of the hot season, and there had not been a drop of rain in the three months since the students had left Maekung.

It had been a long, hard few months. After Ned had left, each day had stretched out, blank and boring. There was little for Jinda to do, no one to talk with, nothing to look forward to. With only her grandmother and little brother left at home, the place seemed empty and silent. The only bright spots had been the erratic visits by the postman, who would putter into their village on a dusty motorcycle to deliver letters from Ned.

When Ned's first letter came, dozens of villagers crowded around as Jinda tore open the envelope. Knowing that she was expected to read it aloud to them, she felt relieved rather than disappointed that the letter was quite impersonal.

It described the hectic organising that he had been involved in, as soon as they got back to Bangkok; the printing and distributing of leaflets, the protest marches, the petitions, the numerous articles in the newspapers focusing on the tenant farmers of Thailand. All this activity seemed rather remote to Jinda.

Ned's letters arrived regularly, and Jinda dutifully read each of them out to whomever had gathered to listen. Only she realised, reading between the lines expressing his concern for Inthorn or frustration at being back in Bangkok, that he cared for her, and missed her.

Towards the end of March, he wrote that he was making plans for a huge demonstration around the issue of the high land rent, and that he hoped that Jinda could come down to Bangkok to speak at the rally then.

It was only in a postscript that he had added, 'Come to give the speech, but come for that other reason too.' Jinda smiled, and did not read that comment out to her neighbours.

In the letter after that, he had included a train ticket and a detailed map of how to get to Sri's house. 'The rally is scheduled for next month,' he wrote. 'I still think a speech given by you, on Inthorn's behalf, would have great impact. Come, please come. You can stay with Sri's family, and I'm sending a round-trip ticket. So be sure to tell your grandmother.' Jinda could imagine his lips curve into a smile as he added, 'that you'll be well looked-after, and that you can return to Maekung any time you want.'

Jinda's grandmother, as Ned had anticipated, did not like the idea of her going. Whoever heard of a girl going off to the city by herself, just because some 'young man' (she said it as if she really meant 'dragon') had sent for her? What would Jinda do there? How could anyone be sure Sri's family would really look after her?

Unexpectedly, it was Jinda's sister who came to her defence. Dao argued, and very convincingly, that Jinda could take care of herself, that it was all in an effort to help their father, that both Ned and Sri's family could be trusted, and that no real harm could come from it. 'After all,' she had said, 'didn't he send a return ticket? Jinda can always come home.'

Their grandmother had held the tickets up right next to her eyes, and squinted at them, even though — as Jinda well knew — she couldn't read. Then she had turned them over and smelled them. 'Can you really come back from Bangkok, with these?' she asked Jinda.

'Yes, Granny,' Jinda had said. 'And I will, too.'

Jinda was grateful to Dao for persuading their grandmother to let her go. Several days later, Jinda realised that Dao's support might have been prompted by selfish motives: Dao wanted to live in town now, with Dusit, the rent-collector.

At first Jinda did not believe her.

'But why?' Jinda asked.

'Because Mr. Dusit has asked me to.'

'That's no reason!' Jinda said angrily.

'Well then,' Dao said, 'because I want to.'

'But Dao, it wouldn't be right. Dusit's the one who put Father in prison, how could you . . .'

'Come on, Jinda. Who's to say who got Father put in prison? Mr. Dusit, who had Father arrested for not paying his full rent, or Ned, who incited Father to resist paying the rent? Besides,' she added, 'Mr. Dusit said he'll arrange it with the prison warden so that I can visit Father in prison once in a while, if I live in town.'

'Where would you live?'

'He's already rented a set of rooms for me,' Dao said quietly.

'What about Ghan?'

'What about him? I haven't heard from him for almost two years. The son I bore him is dead, and even his father doesn't really consider us married anymore. I don't think Mau Chom or his wife want me to go on living with them.'

'Dao, you can always live at home.'

Dao looked at Jinda and laughed, a thin, mirthless sound. 'Why should I?' she asked. 'You're going off to your Ned. Why shouldn't I go to Dusit?'

Jinda felt a flash of anger. 'I'm going to Bangkok,' she said, 'because I'm trying to help get Father out of prison.'

'Right,' Dao said, 'and I'm off to Chiengmai to help Father while he's in prison.'

They stared at each other in silence for a long moment. 'All right,' Jinda finally said, 'I'll help talk Granny into it, if that's what you really want.'

Dao took her sister's hand and pressed it. 'Thanks, Jinda,' she said, then sighed. 'Sometimes, though, I don't know what I really want, anymore.'

How smooth her cheeks were, Jinda thought, and how graceful the sweep of her neck. How can someone this lovely, be so unhappy?

And so, despite her own reservations, Jinda nagged at her grandmother, until after several days, the old woman very reluctantly consented.

'The spores of every fern scatter, when a strong wind blows,' the old woman said quietly. 'But the fern leaf remains, to wait for those spores to settle and take root, and grow around it. I'll wait right here, Jinda and Dao, for you to come back. And I pray that your father will too. Someday soon, I know we will be a real family again.'

The western range of the Chiangda mountains had slid past

now, and Jinda thought she could see the familiar hilltop where she had sometimes watched the train snake its way between the valleys. She could image Pinit standing there now, his small hands shading his eyes from the slant of the sunset, watching the train go by.

The image made her sad, and she turned away from the window.

Jinda looked around her. Her section of the train was packed full. Passengers were crammed four deep into seats, while others perched on armrests and some even curled up on newspapers in the aisle.

A young woman was sitting next to Jinda, hugging a plastic airline bag. Out of the corner of her eye, Jinda glanced at her. Her hair had been permed into frizzy curls, and she wore a blouse so sheer that Jinda could see her bra underneath it.

The girl tore open a paper bag, took out a steamed meat bun, and started eating it. She noticed Jinda watching her, and grinned. 'Here, want one?' she asked, her mouth full of pork.

Jinda's mouth watered. The buns smelt of garlic and basil and meat. She had never accepted food from a stranger before, but then, she thought, she had never been in a train before either.

'Thank you,' Jinda smiled, and picked a bun out from the bag thrust at her.

For a while the two girls ate in companionable silence. Jinda savoured each bite, the bread so chewy smooth and the filling salty and fragrant. She had seen such buns for sale, but never had one before. No wonder they cost three baht each!

The other girl had finished, and licked her fingertips delicately before turning to Jinda again. 'Been down to the city before?' she asked. 'No? Wait till you see it — great place! Some of the shops are so big it'd take you days just to walk through them. A whole floor selling nothing but lipstick and eyeshadow and perfumes, can you imagine that?'

Jinda realised that she wasn't expected to answer, but just to listen. This she did with curiosity. After a long description of Bangkok's department stores, the girl told Jinda that she was moving to Bangkok for work. She had started working in town six years ago, and it was time to move on to the city. 'Don't get me wrong, I liked the place I was working at. Regular clients, always paid in full, and never so drunk they'd hit you or anything. And the housemother wasn't bad either. Of course she took half of what we made, but she did keep the police off our backs. It was a good set-up, and I sent my two brothers through school with the money I made.' She stuffed the last of the bun into her mouth and shrugged. 'I really hated to go, but when it's time to move, it's time to move.'

Jinda listened, confused. 'Why?' she asked.

'It's the clients. They get tired of seeing the same faces year after year. Say they might as well stay home with the wife if there're no new faces around.' She shrugged, examining her red fingernails. 'Most girls only last two or three years in a house. I guess I was better than most of them.'

Jinda was still mystified. 'What sort of work do you do?' she asked shyly.

The other girl raised her eyebrows. 'You serious?' she retorted. 'Come on, what do I look like — a nun?'

With a flash of comprehension, Jinda understood. She flushed.

'Don't worry, I don't bite!' the other girl laughed. 'And you don't have to look so shocked, either. I'm sure girls from your village have landed up doing the same thing. It's always the same story. Father has to sell off the land, the Mother pawns her earrings, and then Daughter pitches in by selling the only thing she has to sell — her body. It's one way to make a living. And it's better than selling off land

and jewelry — after all, you can sell your body again and again.' She nudged Jinda gently. 'Right?'

Jinda nodded, but the bread stuck in her throat. She swallowed it with difficulty, then quietly slipped the rest of the pork bun underneath the seat.

The other girl had taken off her sandals, and now tucked her legs under her. Squirming into a comfortable position, she curled up and went to sleep.

Jinda looked out of the window. There was nothing familiar about the scenery now. The mountain ranges near home had receded from view long ago, and between occasional foothills the fields stretched out, flat and brown. It was dusk, and sprinkled on the horizon were the flickering lights of scattered villages. Jinda felt strange, seeing the darkness creeping in without the familiar sounds of crickets and bullfrogs that usually accompanied it. She felt a pang of homesickness, but brushed it aside. The train had a soothing rhythm that comforted Jinda and soon lulled her into sleep.

She awoke with a start. She thought she had heard little Pinit calling her, his voice shrill with excitement. Blinking, she looked up. A boy was standing in the doorway of the train, hanging on precariously to the handrails as he craned his neck to look out.

'Someone should pull him in before he falls off,' Jinda thought groggily. But it was not yet daybreak, and most of the passengers were still asleep. Getting up from her seat, Jinda climbed over the legs of the girl sleeping beside her, and went lurching down the aisle towards the boy.

'Hey, get back in here,' she called as she approached him.

He glanced at her, then grinned. 'Look!' he said to her.

Jinda leaned out and looked.

Incredibly, overnight, the world had been transformed into a patchwork of lush green. In the first rays of the morning light, newly transplanted rice seedlings glowed a translucent gold. Sparkling water flowed from one paddy field to another, catching bits of sunlight. So much water in the dry season? And how could the rice have grown so tall when the monsoons were still months away? Jinda felt she must have entered a dream world.

In the distance, a little girl in a yellow sarong was skipping along, brandishing a twig at a flock of ducklings. As the train whipped past, she looked up and waved gaily, while the ducklings splashed around her, nibbling at the clusters of duckweed floating between the rice seedlings. I must be dreaming this, Jinda thought. It's too lovely.

'Isn't that something?' a gruff voice said behind her. Jinda spun around, nearly losing her balance on the doorstep.

The man reached out and steadied her by the elbow. 'Careful,' he said, 'or you'll fall out, like I thought my son would,'

Jinda saw that he was holding the little boy's hand firmly in his own. She smiled at them both. 'I can see why you were hanging out so far,' she said to the boy, 'It's really beautiful out there.'

'We're in the Central Plains now, nearing Bangkok,' the father said, gazing out of the door himself. 'The fields are all irrigated here, so they can plant two, three crops a year.'

'No wonder it's so green even now, then,' Jinda said.

'Everything's richer near Bangkok, that's why I'm moving my family there.' He gestured to a group of children nestled against a woman in a nearby seat. Obviously from the countryside, their clothes were faded but clean, their belongings tied up in cloth bundles.

'What're you going to do in the City?' Jinda asked.

'Why, get a job, of course,' he said. 'I've heard construction jobs pay pretty well.'

'And your family?'

'The wife can work a sewing machine — she could be a seamstress, easy. And the boys can go to school, maybe all the way to high school so they can work in air-conditioned banks when they grow up. And we'd maybe run a small coffee shop, sell roast chicken and papaya salad . . .' There was a dreamy look in his eyes as he gazed at the lush fields sliding by. 'And when we have enough money saved up, maybe we'll buy us a big piece of land back home.'

'Where's home?' Jinda asked.

The farmer hesitated. 'It was in Chiengrai. We rented six rai's of land there, but with the harvest so poor these last two years, there just wasn't enough rice to live on.'

'After giving half of the rice crop to the landlord,' Jinda said.

The farmer looked at her in surprise. 'Of course,' he said.

'Of course,' Jinda echoed bitterly.

The train was approaching the outskirts of the city now, and Jinda saw long lines of trucks driving by, piled high with baskets of cabbages, gunny sacks of rice, mounds of pomelos and tangerines. It seemed as if all the produce of these lush fields were being siphoned into the city.

'No wonder they call it the City of Angels,' the farmer said softly, pointing at the trucks. 'There's so much here, it must be like living in Heaven.'

They passed busy streets lined with store fronts: huge glass windows displaying dresses, food, gleaming motorcycles, even sinks and toilets. Jinda grew dizzy just trying to see what was in each window.

The other passengers in the train were stirring now, and people peered out of the windows with sleepy awe. There was a sense of exhiliration among them, as mothers hugged laughing children and men gestured at the scenes outside. A new life, the train seemed to be saying reassuringly, a new life, a good life.

The little boy ran back to his mother, shouting, 'We're there! We're there!'

'Where?' his mother asked.

'The . . . the City of Angels!'

Jinda smiled.

Just then the train flashed past a dense, ugly section of huts. Bits of cardboard, warped planks, plastic sheets and chicken wire were all thrown together as if a storm had pushed all the houses of a very tacky village into one messy pile.

Jinda caught a glimpse of two girls squatting on a lopsided porch, brushing their teeth, and spitting the white foam onto a swamp so putrid the water looked like tar. Jinda blinked twice before the cluster of huts flashed by.

The scene was replaced then by wide streets teeming with cars again. Had she imagined that sudden unpleasant scene, Jinda wondered? Surely nothing like that could exist right in the middle of this City of Angels?

Before she had the chance to mull over the question, the train was pulling into Hualongporn Station.

Jinda squeezed her way back to the seat, and gathered up her bags. Then she joined the rush of passengers pushing their way down the steps to the platform.

Vendors waving trays of sliced watermelon, plastic toys, thick newspapers, swarmed around.

'Want a taxi?'

'Chewing gum! Cough Drops! Tiger Balm! Horlicks! Sour plums!'

'Headline: Singer Mali shoots lover inside Patpong bar!'

Clutching her bags, Jinda stumbled towards the main gate. People seemed bent on blocking her way, trying to sell her things. Passengers from the first and second class compartments barged by, their luggage piled onto flat carts. Jinda eased her way out of the mainstream of traffic, and took a deep breath.

At the edge of the station platform, was a man asleep on a bench, his wife and brood of children huddled on the ground around him. They looked as if they had been camping on this spot for days, their blankets grimy with soot, their faces blank with exhaustion. Jinda stared. This was no trick of the imagination.

Slowly, Jinda walked out of the gate. On patches of lawn, or around a bench just outside the station, were little families huddled together. With a sinking feeling, Jinda realised that these were farmers like the one she'd just met on the train, come down to make a living in the City of Angels.

Jinda took a long hard look around her. The roar of the buses and trucks was overpowering. A veil of dust and smoke hung in the air, so that even the early morning sunlight looked grey. On the steps of the station several children, thinner and more ragged than her brother, Pinit, were selling jasmine garlands, the slender blossoms looking already wilted.

So this, Jinda thought, looking around her, this was the City of Angels.

Chapter 10

The bus ride to Sri's house was long and confusing. Jinda got on the wrong bus twice, and each time she asked the conductor for directions, she was acutely aware of her rural accent. The other passengers on the bus seemed to stare at her in smiling contempt.

When she finally got off at the right street, she was pleasantly surprised to find it so quiet. A shady lane lined with swaying casuarina trees stretched out in front of her, and the air was clear and fresh.

Clutching the letter with Sri's address on it, she walked down the lane. High walls crowned with bits of broken glass bordered the road, behind which tiled rooftops hovered in the distance. Jinda felt as if she had stepped into a world of effortless peace and balance, faraway from the grimy bustle of the city.

The smell of curry drew her further down the lane. She had eaten nothing except that steamed pork bun on the train, and she was ravenous now. She quickened her pace, and round the bend in the road was a small table, laden with tureens of steaming rice and curry.

A plump young woman guarded the table, ladle in hand.

Surely she's not selling the food, Jinda thought. Hardly anybody comes this way.

Just then five monks padded towards her in single file, their rubber sandals slapping softly against the asphalt. Heads bent, arms cradling a round alms bowl, they paused by the table. As the plump woman reverently ladled food into each bowl, each monk, murmured a blessing, and then moved on.

Jinda moved closer. What generous helpings, she marvelled! Eggplant curry, and shrimp paste to go with the roasted fish. In Maekung, half a salted duck egg was considered a delicacy. The people in this mansion must be earning a lot of spiritual merit offering so much food to monks, Jinda thought. They'll probably be born as even more important people in their next lives.

She glanced up at the gleaming name plate on the gate, and with a start saw the number on it. It was Sri's home address. Did that meek little student, Jinda thought incredulously, really live in that huge mansion behind that gate?

The last of the monks padded off, passing Jinda with downcast eyes. Jinda inched towards the food-laden table. Mounds of steaming rice were still piled on one platter, and the bowl of curry was barely half gone. She stood to one side, and cleared her throat.

The plump woman ignored her, and started packing away the offerings. When she clamped a lid onto the pot of eggplant curry, Jinda coughed again, more loudly.

'Go away,' the plump woman said, waving her hand at Jinda. 'This food's only for monks.'

Jinda swallowed hard. Her stomach was churning, and her mouth watered at the sight of the curry. 'But I'm not looking for food,' she said stiffly. 'I have a letter here. See? I'm looking for Number 25, Pratibat Road.'

The plump woman lifted her eyebrows. 'This is the house. What do you want?'

'I'd like to see Srichandra Pramatinodh,' she said, pronouncing the last name carefully.

'You know the Young Mistress?'

Jinda nodded. 'She knows that I'm hoping to live here,' she said.

'Well,' the woman said dubiously, 'You look a bit young, and your accent's pretty thick, but I suppose you'll do. You've picked a good time to come, I must say. Both housemaids have just quit — that's five in three months — and Madame's pretty desperate.' She packed up the trays on the table, and nudged the metal gate open with her foot. 'Come on in, and bring the rice with you. Might as well start making yourself useful.'

Jinda followed her in, through the steel gates and into a huge garden shaded by large tamarind and frangi-pani trees. Rose bushes bordered the whole lawn, and as Jinda walked past them, she felt as if she was being swept away by the scent of roses.

The house at the end of the curved driveway looked like a temple, with huge white pillars looming in front of it. A flight of marble steps led up to a massive door of carved teak wood. Jinda had never seen anything so magnificent in her life.

The plump old woman led her past the mansion, then past a shed where three gleaming cars were parked side by side, and finally to a row of neat, whitewashed rooms. So this was where Sri lived, Jinda thought. No wonder she found our village poor and dirty.

'Which . . . which room is Sri's?' Jinda asked.

The fat woman stared at her. 'Don't be silly,' she snapped. 'This is the servants' quarters.'

She pointed to a bench alongside one wall, and told Jinda she could wait there. As an afterthought, she thrust a plate of food into her hands.

Jinda stared at the plate of steaming rice before her. It smelled of basil and peppers and fried chicken. For a moment she was mesmerized. She had never smelt such a rich mix of meat and spices before. Her stomach growled, and she looked up at the cook guiltily.

'You're hungry,' the woman said flatly, 'so eat.' Jinda hesitated only for a fraction of a second. Then, heart pounding, she spooned up a piece of pork. It was delicious. Barely taking time out to breathe, she ate spoon after spoonful of the white rice, moistened with gravy.

The fat cook lowered herself next to Jinda on the bench, and sighed. 'Does me good to see someone eating like that,' she said. 'The Young Mistress now, no matter what I cook, she hardly touches her food.'

Jinda spotted another morsel of chicken, and scooped it up.

'Not that she even eats at home much these days,' the cook said. 'Just a cup of tea in the mornings, and not a thing until she comes back late at night.' The woman mopped her head with a dish towel, and sighed. 'Her mother's worried sick, poor lady. It's worse than that hairy white boy friend Miss Sri had visiting some years back, after her trip abroad. This year she vacationed in some village up North, and . . .' she shook her head. 'Like a disease, it is, but worse. Know what's hit her?' She leaned towards Jinda and whispered loudly, 'Politics.'

Jinda shovelled the last spoonful of rice into her mouth, and grinned. 'I know,' she said. 'It's contagious, and I think she caught it in my village.'

The cook glanced at Jinda sharply, but before she had time to say anything, a clear, high voice called. 'Noi! Noi, where are you?'

The plump woman jumped up with surprising speed for one so heavy.

'Noi? Or Lek then, Lek! Lek! Where are any of you?'

The backdoor of the mansion swung open, and a tall thin woman stepped out, wrapped in a gauzy, floor-length robe. 'Somboon!' she called to the cook. 'Where's Noi? or Lek?'

'Noi left your service, Madame. Last week, if you'll remember.'

'And the other one? Lek?'

'Yesterday afternoon,' Somboon said.

'But she can't have. She knew I was having an important guest today. She must have planned this.'

'You fired her,' Somboon pointed out quietly.

'Did I? I suppose I did. Still, it's very inconvenient. What about the new maid. Wasn't she supposed to come this morning?'

'Yes, but . . .'

For the first time, the tall lady seemed to notice Jinda. 'You,' she said peremptorily. 'What's your name?'

'Jinda.'

'Say, "Madame"!' Somboon hissed in her ear.

'Madame,' Jinda echoed obediently.

'Have you worked as a serving maid before?' she asked.

'No,' Jinda said, then added, 'Madame.'

'I wish we had more time to break you in, but I suppose you'll have to do for now. Somboon, see that she puts on some decent clothes, and show her where the tea-things are. We'll use the Celadon set today. And the strawberry tarts. If the living room doesn't have fresh flowers in it yet, we'll have to use the morning room.'

She walked off, the hem of her long robe sweeping after her, leaving Jinda in a daze.

'Come on, this is your chance. Don't look so worried, it's not all that hard,' Somboon said. 'I'll show you what to do.'

'Show me what to do?' Jinda repeated dumbly. 'Just what am I supposed to do?'

'You heard Madame. Serve tea to whoever's coming, of course. Hurry up and change, you don't have much time.'

Jinda took a deep breath, and crept into the room. The marble floor was cold underneath her bare feet, and she felt stiff and clumsy in the starched white blouse and black sarong she was wearing. In her hands she held a tray on which Somboon had arranged the tea set and pastries.

'Lower!' Somboon whispered behind her, as Jinda took a few steps into the room.

Jinda stooped down even lower, so that she was almost doubled over. Not even for the village monks had she bent this low, Jinda thought resentfully, why did she have to stoop so far for a few rich old women?

As Jinda approached the coffee table, she saw three pairs of women's legs, each neatly crossed and in glossy high-heeled shoes. She remembered Somboon's hurried instructions, as she set the tray onto the table. 'Get on your knees as you get to the coffee table, and serve kneeling.'

Carefully Jinda knelt down by the low table, and set each teacup on it. Then, she started pouring the tea. A few drops of water splashed out onto the table.

'Gently,' she heard the lady murmur, as Jinda passed it forward. A pair of bare, pale arms reached out and took it.

Jinda was too engrossed pouring the second cup of tea to notice that someone else had just come into the room.

'I'm awfully sorry to interrupt, Mother,' a familiar voice said softly, above Jinda's head.

'What is it, dear?'

'I just wanted to let you know I won't be home till late again tonight. I have a meeting at Thammasart.'

With a shock, Jinda realised that it was Sri speaking. Her accent and diction were so different — so much more fluent and assured — that it almost sounded like another person.

'How late will you be?'

'Past midnight, I'm afraid.'

'Take your car, then.'

'But mother, please. The buses are more convenient . . .'

'More "proletarian", you mean.' Sri's mother nodded towards the other ladies. 'You see how quickly I've picked up their jargon,' she said mockingly.

There were polite laughs.

Jinda finished pouring the tea, and held the teacup out, wondering whether to set it down or hand it to Sri's mother.

'Here, sit down for a minute and join us for a cup of tea,' Sri's mother said, motioning Jinda to serve the tea to Sri.

Hesitantly, Jinda held out the steaming teacup to Sri. The student ignored it.

'Mother, I'm late,' Sri said. 'I'm not having any t . . . tea with you, and I'm n . . . not taking the car!' her voice was tense, and the slight stutter Jinda knew so well was back in it.

'Fine, don't have any tea,' said her mother evenly, 'but you will take your car. I will not have you wandering alone all over the streets at night in taxis and buses.'

'Mother, I . . . I will n . . . not take the car.'

'Then, Sri, you will be back tonight by nine o'clock.'

'Nine o'clock!'

'That's when the women's dormitories close, isn't it? If you're not going to observe the rules I've set up, you can move right back into those awful little dorms again.'

'But, Mother . . .'

'Goodbye, dear. Have a nice day at the revolution.'

Sri turned so sharply that she brushed against the low table, and knocked the teacup onto the floor. The hot tea splashed onto Jinda, scalding her hand. Jinda cried out.

'Jinda!' Sri dropped down on her knees too, and with the hem of her skirt, was wiping off Jinda's arm.

'You know this girl?'

'Know her? Mother, she's Jinda. I lived with her family for two months. She's like a sister! You . . . you can't treat her like a servant. Kneeling like that. Come, Jinda.' Sri scrambled to her feet, tugging at Jinda.

Confused, Jinda allowed herself to be pulled up.

'Wait a minute,' Sri's mother said. 'At least pick up the broken porcelain. It is my best set, after all.'

Automatically, Jinda dropped to her knees again.

But Sri would not let go of her arm, and almost angrily

yanked Jinda back up again. 'Don't let her order you around, too,' Sri said vehemently. 'Just because she's used to being obeyed, she thinks she can oppress every . . .'

'That's enough, Sri,' her mother's mocking tone had turned steely. 'Either join us for tea, or go and make your revolution. I hardly think you can do both at the same time.'

'Don't worry, I'm going,' Sri said, blindly pulling Jinda with her.

'And please,' her mother called after her. 'Don't bring any more stray "proletarians" home with you anymore, will you?'

On the bench where Jinda had been eating just now, Sri sank down and drew deep, shuddering breaths. Her face was flushed a deep red, and her eyes looked suspiciously shiny. Jinda sat down beside her, and squeezed her arm.

'She has no right, no right,' Sri stammered, 'to treat you like that. To mock, and insult you . . .'

'She didn't know, Sri. She thought I was the new maid. It's all right, I don't mind.'

'But it's like this all the time, the deliberate humiliation, the mocking . . .'

Jinda realised with a flash of insight that Sri was referring more to herself than to Jinda. She took her friend's tightly clenched fist, and gently prised it open. Sri's hand was so pale and smooth, compared to Jinda's own tanned, calloused one. How strange, Jinda thought, as she patted Sri's hand gently, how strange this is. Sri has everything, money, education, a home, — and I have nothing. Yet here I am, trying to comfort her.

'It isn't your fault, Sri, don't feel bad. And don't worry about me, I can always stay at Ned's place.'

Sri lifted her tear-stained face, and Jinda wiped it with the edge of her sleeve.

Just then Jinda caught a glimpse of the cook's face, as she peered through the kitchen window at the two of them.

Somboon looked so shocked that Jinda couldn't help but burst out laughing.

Little muddy canals wove alongside dirty streets, clogged with rotting debris. Jinda looked down from the bridge over one canal, and watched a watermelon rind float past, hooked into the handle of a broken cup. Then a dead cat bobbed by, its bloated belly humped into the water. For a moment it was caught and entangled to a clump of scraggly ferns, where it gazed at Jinda with glassy eyes and a set, sharp-toothed grin. Jinda walked away quickly.

For the third time she checked the return address on Ned's last letter. Ned lived, Jinda was discovering after several hours of wandering, in a different neighbourhood from Sri's.

Down a line of storefronts, Jinda passed a bald manne-quin, its plastic breasts jutting out above a sequined skirt. In the shadows behind the mannequin pale seamstresses slouched over their sewing machines, their faces shrouded in swathes of paisley cloth.

Down the alley, narrow shacks were crowded together, each one tacked together from sheets of plastic and jute bags and corrugated metal roofing. Children swarmed in the tiny spaces between these houses, tossing bottlecaps and mud marbles in intricate games. Jinda threaded her way through them, stepping on the zigzag of planks which served as walkways above the stagnant, mosquito-ridden swamp.

She passed by an old wooden desk under a tall banana tree. Behind it a schoolboy in uniform was bent over his textbooks, one hand stroking a plump hen nesting in the top drawer. The little boy looked up as Jinda passed, and smiled. 'Where're you heading, sister?' he asked conver-sationally. It was the first time anyone had said anything pleasant to her all day, and in the familiar Northern accent too!

'Looking for a friend,' Jinda replied, stroking the hen as well.

'Anybody I know?'

Jinda showed him Ned's envelope.

'Brother Ned!' he exclaimed, smiling at her even more warmly. 'Everybody knows him. Why, he's famous. His picture was all over the newspapers last year. He was even on TV once.'

Jinda grew uneasy, listening to this. She had known that Ned was a student leader of some sort, and that the three students who came to Maekung with him certainly deferred to him, but this was the first inkling she'd had that he was 'famous'. Jinda looked dubiously at the schoolboy. He's only a child, she told herself, he's exaggerating. Still, her experience at Sri's house that morning had shaken her more than she was prepared to admit. After all, if Sri in Bangkok was so different from the Sri in Maekung, what would Ned in Bangkok be like? Even more fluent, perhaps, and more intellectual? Jinda hoped not.

'He's just down the lane here,' the schoolboy was saying. 'Shares a house with a few other students. My uncle's his landlord, but doesn't charge him much rent, because he admires Brother Ned so much. He's really great, Brother Ned is!'

Her heart beating fast, Jinda thanked him and walked on. She was nearing his place at last.

It was late afternoon, and Jinda felt strangely at home here, as if she was back in the village. Mothers smeared white rice-paste on the faces of their freshly bathed toddlers, old men watered pots of orchids or lemon grass while boys showered in the open air, their heads frothy with white soapsuds. These people are all from the countryside, Jinda realised, like me. It was as if they had built a sprawling village in the middle of the city.

Ned's house was a small wooden structure, two-storied but so narrow that the laundry from the adjoining house protruded through its front window. Its wooden shutters creaked on rusty hinges as a stray breeze blew through.

As Jinda approached the house, she heard sharp yipping cries. Curious, she walked towards the noise.

She saw a pack of fifteen or more puppies, all crowded around a washbasin, yapping and slurping. Most of the puppies were mangy and thin, but they all wagged their tails furiously as they fought for a niche at the feeding trough. Some man with his back towards her was busy pouring a pail of leftover food into the basin.

Jinda smiled, watching them. The scene reminded her of her father feeding piglets after a bountiful harvest.

Then, pandemonium broke out when a large mongrel ran into the crowd of puppies, trying to get at the food. The man with the slop bucket nearly toppled over; the big dog snarled and snapped, and the puppies howled.

Jinda looked around, and saw a stick lying near by. Snatching it up, she ran over and swung it into the dogfight, trying to help break it up. She managed to prise the big dog away from the puppies.

'Grab its tail!' she shouted.

'Got it!' the man cried, grabbing the mongrel by its tail. 'Ned!'

'Good Lord, Jinda!'

Ned let go of the dog the same instant that Jinda dropped her stick. And the dogfight started all over again.

Laughing and cursing, Ned bent down and yanked the mongrel out again, tugging at its hind legs. Jinda chased it off as Ned calmed the puppies down, and poured more slop into their wash-basin.

When they finally faced each other, Jinda's sarong was torn, and her hair, which she had pinned up especially carefully in anticipation of seeing Ned, was loose and dishevelled. Ned looked even more messy, with his clothes and face splattered with mud.

They stared at each other, and burst out laughing. Jinda

laughed so hard she doubled over, and had to be guided, stumbling and laughing, onto the porch in front of Ned's house. The famous student leader, Jinda thought with delight and relief, the glib intellectual — breaking up a dogfight!

Ned's face was lit up in a huge smile. 'I wasn't sure if you were ever coming,' he said. 'I didn't want to be too insistent, but you never wrote . . .'

'I did once,' Jinda said, 'and you corrected my spelling in your next letter.'

'Is that why you never wrote again? God, I'm so stupid sometimes. I'm sorry I made you angry, Jinda.'

'Never mind,' she said, deciding not to tell him that she had been ashamed, not angry, by his corrections.

'When did you get here?' Ned was asking. 'Have you been to Sri's house yet?'

'Yes,' Jinda said, and told him briefly what had happened there.

'Poor Sri,' he said when Jinda had finished. 'She moved out of the dorms because she didn't have enough freedom there, but things don't sound much better for her at home.' He squeezed Jinda's hand lightly, and smiled. 'Well, you're very welcome to stay here, Jinda. Come on in, I'll show you where you can put your bags.'

Jinda hesitated, and Ned immediately understood. 'Don't worry, little girl. I won't do a thing that your grandmother might frown upon!' Jinda relaxed then, and laughing, followed him into the house.

It was cool and dim inside. Jinda looked around her. The room was bare except for piles of newspapers stacked high in one corner, and a large desk by an open window. On the wall were tacked large posters of strange looking men who seemed to glare down at Jinda as she entered the room, two of whom were bearded, and the third a benign looking old man with a high forehead. Quietly Jinda spelt out their names: Karl Marx, Vladimir Lenin, Maotsetung. 'What strange names!' she murmured.

Ned pointed to another poster tacked right above his desk. 'Not this one. He's Thai.' Jinda studied the face with interest. It was of a slender young man, with wire-rim glasses, a cigarette dangling between his thin lips, his longish hair dishevelled. 'Jit Pumisak,' Ned said. 'The first revolutionary of Thailand. He helped translate the writings of those other men on the wall, and he also spent years living in villages in the Northeast. He's the one who wrote the book on Thai feudalism and . . .' Ned stopped himself, and smiled. 'Go on, tell me I'm talking like a book again,' he said.

Jinda shook her head. 'That's all right. It's nice just being here . . .' with you, she almost said, and bit her tongue.

'But you must be tired. That long train-ride, then Sri's dragon-lady of a mother, and finally . . .'

'A dogfight!' Jinda finished for him. 'You're right, I guess I would like to rest. And a shower,' she added, looking at her dirty sarong and grimy feet.

Ned led her up a narrow flight of steps at the back of the house, and swung open a door to a little room with an open window. 'Our official guest-room,' he said grandly. 'I keep it for students who're too tired to go home after late-night meetings. But it's all yours now. There're blankets and bedding in the corner.' He paused at the door and said, 'If you want to shower, the bathroom is downstairs, on your left. Take your time, I'll wait for you.'

In the little bathroom, Jinda peeled off her clothes, and dipped a ladle into a large earthenware urn of water just like the kind they had at home. She splashed the water over herself, and scrubbed away the day's grime. The water was clear and chilly, and by the time she dried herself and wrapped a fresh sarong around her, she was feeling completely refreshed.

She saw Ned standing in the main room, reading by the window. It was dusk, and the sunlight was already fading, so that he had to squint a little. Brushing off the beads of water from her bare arms and shoulders, Jinda went over to his desk, and turned on the reading light there.

'You shouldn't read in that dim light,' she said.

'Thanks,' Ned said, looking up to smile at her. His smile suddenly faded and he turned away abruptly, so that he stood facing the poster of Karl Marx. 'And you shouldn't walk around in just a sarong. 'You're not in the village now, and I don't think your grandmother would approve.'

Jinda tried to suppress her laugh, but couldn't. 'For a moment there, I thought you were talking to that white-bearded Uncle Karl Marx!' she spluttered. Then, still giggling, she ran up the stairs to her little room.

Jinda had almost finished dressing when there was a knock on her door. 'What is it?' she called.

'I have a meeting in a little while,' Ned said through the door. 'Would you like to come?'

'Would it take very long?'

'I'm afraid so,' Ned said. 'Probably past midnight.'

Briefly, Jinda wondered if it was the same meeting Sri had had planned to attend. How many meetings did these students go to anyway, she wondered. 'I think I'll just stay here,' Jinda said. 'I feel pretty tired.'

'All right, and if you're hungry, there's some fruit downstairs.'

After he left, Jinda stared at the closed door. For a moment she felt alone, and a little depressed. She wondered whether her grandmother would have trouble starting the cooking fire without her, and she hoped Pinit would see to it that the drinking jars were filled with well-water. For the first time, she thought she understood a little of what her father must feel, locked up in prison and cut off from his family and the land he loved. But there's no sense brooding, Jinda told herself. She was here to help get him released from prison, and she would try everything she could to do that.

Jinda spread a rattan mat out next to the window, then

leaned out. The window overlooked the front door, where a group of Ned's stray puppies slept, nestled together for warmth. Jinda smiled. Feeling a little like a stray herself, she pulled out a cotton shawl from her bag, and bedded down under it. The shawl smelt of home, of charcoal smoke and straw, and, comforted by that, she snuggled under it. The noises of the street languished as night fell, and Jinda soon dropped off to sleep.

She was awakened by a sound outside the door. Footsteps? She listened again, but the whole house was quiet. It had turned cool, and a fresh breeze blew in through the window. The meeting, Jinda thought sleepily, must have broken up at last.

The door to her room swung open noiselessly, and someone tiptoed in. In the dim light from the open window, Jinda saw that it was Ned.

Her pulse quickened. Peering at him between half-closed lids, she lay very still, pretending to be asleep. He walked into the room and stood over her for a moment, then knelt down by the foot of her mat.

There was a faint thud as he set something down on the floorboards, then the flare of a match. Shielding the flame with one hand, he bent over and lit something. After he had blown the flame out, Jinda saw a faint red glow, smaller than the tip of her grandmother's cigar, near her feet.

He edged over to Jinda, and slowly reached out an arm towards her. Jinda held her breath. Gently, taking care not to wake her, Ned lifted a corner of the shawl and pulled it up over her bare arms.

Then he stood up and slipped out of the room, closing the door behind him. After he left, Jinda sat up to see what the glow by her feet was. Balanced on the neck of a Singha beer bottle was a mosquito coil, its musky incense curling up to keep the insects at bay. For a long time, Jinda watched the tendrils of smoke spiralling up. So he still cares, after all, she thought.

Smiling, Jinda lay down again and closed her eyes. This time she slept soundly through the night.

Chapter 11

The next morning, Ned took Jinda with him to Thammasart University. At first Jinda felt rather intimidated. The campus teemed with students, all clutching thick books and hurrying from building to tall building. Many of them were in student uniform, wearing starched white shirts, and black trousers or skirts. What impressed Jinda most were the tiny steel buttons and shiny belt buckles each student had, embossed with — Ned explained laughingly — the department they were in. 'Pure snob appeal,' he said, apparently by way of explaining too, why he and most of his friends pointedly rejected their uniform, and instead wore the dark blue workshirt and baggy trousers that farmers wore in the village. Not a single girl, however, wore a sarong, and Jinda felt very old-fashioned and out-of-place in hers.

In Ned's classes, even though her sarong was hidden under the desk, she felt self-conscious again. She was the only one in the classroom who did not have a pen and notebook in front of her, and who wasn't busily scribbling down notes as the professor lectured. Even if I had ten pens, Jinda thought wryly, I wouldn't be writing anything down, since I can't understand a thing he's saying. He strings those long words together as quickly as cousin Mali strings her garlands of jasmine buds!

A small chubby man with a moustache, the lecturer brandished a piece of chalk at his class and talked, as far as Jinda could make out, about turning society upside down. 'He's just returned from America,' Ned whispered to Jinda as the professor paused for breath. 'He's read Marx, Lenin and all of Mao, and joined massive anti-war demonstrations in Washington: a real progressive!'

Jinda nodded, but she was not impressed. What did any of that have to do with the rent issue back in little Maekung? Why did he talk so much of fighting and violence, when he looked as if he'd cringe at any village tough waving a broken beer bottle? Jinda shook her head. City life was a vast, confusing web. It was no use trying to understand it.

After class, Ned led her over to a group of students standing in the corridor, and introduced her as Inthorn Boonrueng's daughter. 'You know, Inthorn of the rent resistance movement,' he added.

This seemed to excite the students. They clustered around her, and asked breathless, complicated questions about landholdings, about the farmers' resistance movement, about her own political position. Bewildered, Jinda backed away. 'Hold on,' Ned said, waving aside their barrage of questions. 'Right now we've got more important concerns than mere politics. Jinda and I are going to have lunch. Anyone want to join us?'

'This will give you a chance to meet some of the people you'll be working with on the land reform rally,' Ned told Jinda quietly as he steered her away. 'Some of them are textbook radicals — they talk progressive, but they haven't set foot outside the classroom — but don't worry, they all mean well.'

Jinda glanced back at the students following her, chattering and hugging their books to them. Sure, they mean well, she thought, but can they get Father out of prison?

At the cafeteria, Jinda stood in front of the long line of food, gazing at tray after tray piled high with curry chicken, vegetables and meat. Stunned at the sight of so much food, she stood before the tray of fish curry and another of stirfried cabbage, torn between the two. Finally, feeling like a glutton, she had asked for both. Over her mound of steaming rice, two ladles of curry and meat had been spilled. Jinda was overwhelmed. It was not until she had sat down with Ned and a group of his friends that she realised they had each taken three, even four, different courses!

Did city people eat like this every day? No wonder Ned and Sri had thought the villagers' daily fare of broken rice and fish sauce inadequate, Jinda thought.

That evening, when they got back to the house, there was already a neat semi-circle of shoes around Ned's doorstep. A steady drone of voices reached them through the screen door. 'We're late,' Ned said. 'The meeting's started without us.' He kicked off his sandals, and went in.

Jinda remained outside, looking at the shoes. White canvas tennis shoes, most of them, a few pairs of leather shoes, some rubber slippers, even one pair of shiny high-heels. Slowly Jinda kicked off her own rubber sandals. With quiet satisfaction, she noted that hers were worn and rust-brown with dirt, the only pair to show any sign of contact with the soil.

In the next few days, Ned brought Jinda to so many places and introduced her to so many people that her head spun. She followed him to newspaper offices where she was interviewed by people who noted down everything she said. She visited a textile factory where workers were on strike. She made the rounds of other schools active in organising the upcoming rally. She met farmer leaders from North and Northeastern Thailand who had come to Bangkok to speak up against the high land rents.

Jinda came to spend most of her time at 'The Head-quarters', a set of dingy rooms in a two-storey building near Ned's house, across the river from Thammasart University. There, students who had been elected to represent each of the dozens of universities and vocational colleges all over Thailand, gathered. It was a tight-knit network, capable of mobilising thousands of students into mass rallies at a day's notice. And Ned, as one of the representatives of Thammasart University, was pivotal in this network.

At the Headquarters, on dilapidated desks piled high with leaflets, earnest looking students banged away incessantly at old typewriters. A battered mimeograph machine churned out yet more leaflets, with a constant soft staccato stutter. Ceiling fans churned the still air, often swirling up a piece of paper or two.

At first Jinda could not get used to the constant whirl of noise and activity at the Headquarters. If she was attending a discussion group on land-reform, she would be distracted by the news report coming from a radio which was never switched off. Or if she was helping to write an article about the farmers' rent movement, the contact prints which the student photographers carried out from their little darkroom next door always seemed more interesting.

But it was at the long political meetings that Jinda found it hardest to concentrate. She felt alternatively bewildered, then bored by the talk, and her attention would wander off. She'd listen to the sound of the children playing in the street below, or watch the noodle-vendor chopping green scallions in his stall. Sometimes, when she got really restless, she would water or prune the potted orchid plant hanging by the window. Its waxy green leaves reminded her of the thick foliage at home, and just touching them always calmed her.

Sri also attended these meetings, and Jinda liked sitting next to her. The pale medical student had a way, Jinda noticed, of asking sensitive questions in a quietly blunt way which reminded Jinda of her own grandmother. And often too, Sri would show signs of impatience just when Jinda herself was feeling restless.

But Sri came to these meetings less and less frequently. To the group she pleaded her busy schedule at the hospital as an excuse, but to Jinda she confided during one meeting that she couldn't get along with Kamol.

'I thought it had something to do with him,' Jinda said sympathetically. She had noticed how this Kamol, a brawny, thick-lipped student representing a big vocational school, often baited Sri about her family's wealth. 'He just talks mean. Don't let him bother you.'

'I wish I could,' Sri murmured, 'but . . .'

'In case you girls haven't noticed,' Kamol broke in loudly, 'this is not a gossip session. We're ready to start the meeting now.'

'See what I mean?' Sri shrugged, and they both sat back quietly.

Ned cleared his throat, and began.

'As you know,' he said, 'the purpose of this meeting is to discuss the increasing violence which has disrupted our last demonstrations.'

Interested, Jinda sat up straighter. This was a topic which she was particularly concerned about.

'That's nothing new,' Kamol said. 'Hooligans and even plainclothes policemen have been throwing smoke bombs and disrupting our rallies for months now.'

'Yes, at that anti-corruption demonstration in March, the smoke bombs which were thrown at us caused a stampede which injured dozens,' another student agreed.

'The difference,' Ned said, 'is that we have never retaliated.' He paused and looked at Kamol. 'This last rally, two policemen were hospitalised for gunshot wounds. Two policemen,' Ned repeated for emphasis.

'So what?' Kamol exclaimed. 'Why should it always be our people getting hurt? It's about time we fought back.' A few of the other students nodded in agreement.

'That's not the way to fight back,' Sri said. 'Violence only creates more violence.'

'Sure, Miss Prissy Pacifist can always hide in her father's mansion,' Kamol taunted, 'while the rest of us scramble from the bombs.'

'That's enough, Kamol,' Ned said quietly. 'I think Sri's right. There are more constructive ways to fight back than using violence. If we start shooting back, it'll only give the opposition the excuse they've been waiting for to really hit out at us.'

'They don't need any excuse,' another student said. 'There're already rumours that gangsters calling themselves the Wild Boars have been recruited and armed by the military specifically to disrupt our demonstrations. They're planning to throw home-made bombs at the big farmers' rally next week.' He sounded nervous.

'That's not all,' Kamol added. 'I've heard that the police patrolling that rally will be fully armed. Not just the usual revolvers, but with bazookas, M-16s, sub-machineguns.'

'How would you know?' Sri asked.

'I heard a couple of soldiers bragging about it one night,' Kamol answered, 'after they'd got good and drunk.'

'Where?'

'Nit's bar,' he said, and smiled. 'I happen to like spending time there. Got any objections?'

If Sri did, she did not voice them.

Again Ned deftly picked up the thread of the discussion. 'The point remains: even if they arm themselves, it is crucial that we keep our demonstrations peaceful. If we resort to violence, we don't stand a chance against them.'

'What are you advocating? That we just stand around at the rally and let ourselves be gunned down one by one?' Kamol sneered. 'Is that your idea of being a hero?'

'Nobody's talking about being a hero,' Ned snapped. 'What's at stake here are crucial issues. The land-rent issue at the next rally, for instance.'

Jinda nodded, grateful that Ned kept the focus on the rally which she had worked so hard to help organise.

'It sounds dangerous,' the nervous student said. 'We know they're getting increasingly violent, yet we refuse to arm ourselves at all,' he shook his head slowly. 'Maybe that next rally should be cancelled.'

'No!' Jinda and Ned said at the same time. He glanced over and gave her a quick smile, then continued more evenly. 'That would be like admitting defeat. We must continue to protest and call for changes in Thailand, but we must also continue to do that non-violently. We've already achieved so much by peaceful means. We must not jeopardize our position.'

'What have we achieved?' Kamol challenged. 'Some token reforms to keep us quiet, that's all.'

'New laws on minimum wage, trade unions, progressive income tax, better hospitals. And now Parliament may even change the land-rent law,' Sri said. 'You call those token reforms?'

'Yes, I do,' Kamol insisted. 'I'm talking about revolution, not reform. I'm talking about drastic change and absolute power, the dictatorship of the proletariat. I'm talking . . .'

You're talking, Jinda said silently, just like that long-winded chubby professor in Ned's class.

'I've had enough,' Sri said vehemently. 'All this talk of power and guns. Everything we do, everything we talk about, seems to come back to Kamol's obsession with power and guns. I thought our purpose was to help the Thai people. What do guns have to do with that?' She was very pale, and her voice trembled.

'I would have thought that it was obvious,' Kamol said. 'This country's heading for a revolution. It's senseless to pretend it isn't, that bits and pieces of reform will improve everything. It's senseless not to arm ourselves and fight back, when they start fighting.' He stared at the grim faces around him, and asked dramatically, 'Are you willing to take up arms and fight, or,' he looked straight at Sri, 'are you too scared?'

Sri stood up very straight. 'I am confident enough of my own courage,' she said quietly, 'without needing to prove it to everyone else. Apparently, Kamol, you aren't. I feel sorry for you.' She turned and walked quickly out of the room.

There was a moment of stunned silence. Then Kamol said, 'There she goes, running off to her father's mansion again!'

Jinda wished she could think of a retort as articulate as Sri's, but since she couldn't, she simply got up and walked out of the room too. Behind her the discussion resumed, a few voices holding forth loudly against a background drone. Jinda was glad to be out of the room.

She caught up with Sri in the alleyway behind the building. The pale student's cheeks were streaked with tears, which she hurriedly wiped away when she saw Jinda.

'I'm not going back,' she said.

'Nobody's asking you to,' Jinda said. 'I just wanted to tell you how proud I am of you, for saying all that,' she smiled, 'without once stuttering!'

Sri smiled too, and took Jinda's hand. Together they walked down the street. 'It's not just Kamol,' she said, addressing her feet. 'They think that just because I'm from a rich family, I don't really care. That I'm not really committed.'

They had reached the wider street beyond the alley, and Sri paused beside a small red car. 'Sometimes I wish I could just forget the politics and go and live in some place like Maekung. You know, I miss the people there, your grand-mother and Pinit especially. It's strange,' Sri said wistfully, 'looking back on it, I think I was really happy there. Things seemed simpler, and more real.'

Sri fished out a set of keys from her briefcase, and proceeded to unlock the door of the car she had been leaning against. 'Listen to me,' she said with a dry laugh, 'talking about returning to your village, when I'm about to drive off in a car which costs more than your family could earn in thirty years. Maybe Kamol is right, Jinda. Maybe I'm just a hypocrite.'

Jinda saw the plaintive look in her friend's eyes, and felt a rush of affection for her. Impulsively, she reached over and hugged Sri lightly. Warm and bony, Sri felt like a small sparrow. 'Don't fret about it, Sri,' she said gently. 'Granny always said there are many roads to the same shrine.'

Sri smiled gratefully at Jinda. 'Thanks, sister, I'll remember that,' she said and got into the car. 'I'm going back to the hospital. At least I know I'm useful there. Do you want a ride? Are you coming my way?'

Sadly, Jinda shook her head. 'No,' she said. 'I'm not going your way.' She stood on the curb, and waited until Sri's car was out of sight, before she turned away.

The next few days were busy ones. Jinda became increasingly involved with the student activity planning the rally on land-rent. She spent most of her time at The Headquarters, and was very much caught up in the discussions there on the problems of land-rent. Large groups of tenant farmers from all over Thailand were convening in Bangkok, in preparation for the rally. Never before had farmers and students worked so closely and with such high spirits for the same goal. She found herself more hopeful than ever that, with the pressure generated by this rally, her father and the other rent resistance leaders would be released from prison.

The only thing that dampened Jinda's spirits was that, as she became busier, Ned seemed more distant. True, they spent more time together now than they ever did in the village, but ironically, they shared fewer private moments together. Jinda remembered the quiet walks they had in Maekung, or the cheerful discussions they had walking to Ned's classes in the first few days after Jinda's arrival in

Bangkok, and wondered why Ned never seemed to find the time to be alone with her anymore.

The night before the big rally, Jinda returned home after a long meeting and found, somewhat to her surprise, that Ned was at his desk — alone. Sitting under the lamplight, with his chin cupped in one hand, he was staring out of the window. He looked, Jinda thought, strangely wistful and worried.

'Ned, what're you doing?' Jinda asked as she came in the door.

He swung around, looking startled. 'I . . . I was just thinking, Jinda,' he mumbled.

'About what? The rally tomorrow?'

Ned shook his head. 'About you,' he said flatly.

Jinda took a deep breath, and waited.

'You've been here almost three weeks now,' Ned said after what seemed like a long pause, 'and you seem to get on well with the people at the Centre, and to enjoy the work there. But,' he looked at her, 'I did send you a roundtrip ticket, after all, and . . .' he frowned, shaking his head slightly, 'well, I was wondering what to do with you, after the rally.'

So that's why he's been avoiding me, Jinda thought. He just wanted to pack me off after the rally. 'You don't have to "do" anything with me,' Jinda said coldly. 'I've decided what to do with myself, thank you.'

'And what's that?' Ned asked quietly.

'To use that train ticket and go right home, of course.'

Ned turned away from her, and stared out of the window again. 'I thought you might,' he said, and his voice was so low she could hardly hear him. Jinda thought his shoulders sagged a little, but that's just because he's tired, she told herself.

'I think that's the right decision, Jinda. I won't worry as much, knowing that you'll be happily settled at home,' Ned said.

If you want to get rid of me, just say so, Jinda thought

bitterly. You don't have to make a pretty little speech about it. 'Fine, I'm going to get some sleep now,' she said abruptly. 'Goodnight.'

'Jinda, wait! Don't go yet. Why don't we . . .' Ned hesitated, 'why don't we work on your speech once more.'

'I'm tired,' Jinda said.

'Just for a while, Jinda. You need the practice. You want it to be perfect for the rally, don't you?'

Jinda managed a thin smile. 'Of course,' she said. She waited as Ned rummaged through the piles of paper on his desk for the notes on the speech.

As she took the papers from him, she noted how carefully he kept his distance even when handing them to her. Quickly, she glanced through the notes, reviewing them. They had worked on it for many hours, weaving together Jinda's memories of her father with the complex issue of land-rent. Ned had insisted that she memorize it rather than rely on notes, and had already spent several sessions coaching her on her delivery of it.

She stood in the middle of the room now, and started her speech.

'Don't clasp your hands behind your back like that,' Ned said, 'You're not reciting a lesson in front of a classroom.'

Guiltily, Jinda unclasped her hands. She had often wondered why people always listened spellbound whenever Ned spoke, but she was beginning to understand that it wasn't just what he said, but how he said it. A flick of the wrist here, a pause there, a sudden unexpected smile, and his listeners were captivated. Practising her own speech over the last week, Jinda had learnt some of his techniques.

Now, as she spoke about the drought in Maekung, she knew how to modulate her voice, when to pause for effect, even how to gesture a little with her hands. Eloquently, she talked of her father's care for his land, of the high rents, of her father's decision to resist paying the traditional half of his crop.

On only one point did she falter. When she started to describe Inthorn when she last saw him, standing behind the iron bars, she stammered and could not go on.

'Talk about his leg irons,' Ned prompted.

'There . . . there were heavy shackles chained to his ankles,' Jinda said with an effort, her voice strained. 'He looked thin and tired, as if he hadn't slept for days. His wounded hand . . .' Jinda swallowed hard, but the ache in her throat would not dissolve. 'I can't,' she whispered.

'You've got to, Jinda. The leg irons are a political symbol . . .'

Jinda's nerves, already frayed, snapped. 'Political!' she cried. 'To you everything is political! My father's a man, don't you understand? He taught me to fly kites, he whittled dolls for me. Is that political? No. So you don't give a damn about that. You only want me to talk . . . to talk about how those awful leg irons — because they're "political", I suppose . . .'

'Calm down, Jinda, you're tired.' He put his hand on her shoulder, but she shook it off.

'I suppose I'm interesting to you only because I'm political too, some political souvenir you've brought back with you from your trip to the countryside!'

'That's not true,' Ned said quietly. 'I care about you.'

'You care about your precious politics!' Jinda snapped, her eyes flashing. Her voice sounded loud and shrill, but she didn't care. 'You wanted me here to help "politicise" people, and after that I should scuttle back to my little village, where I belong. No, you don't care about me. You've never cared!' Horrified at her outburst, she turned to run out of the room.

But Ned caught her by the arm. 'Jinda, calm down. It's not true. Listen to me.'

Jinda took several deep breaths. 'I'm listening,' she said more quietly. But I'm not going to believe you, she thought.

I've heard too many of your speeches. I know how smooth and glib you can sound.

'Jinda, I do care,' Ned said slowly. 'Maybe I haven't showed it, but I do care. I admit, when I first went to Maekung, I thought of it as a political act. I was interested in the issue of land-rents, and I wanted to do something about it. But then, then I got to know you, and . . . to like you, Jinda.' Ned paused uncertainly, 'You told me once, that I talked like a textbook, do you remember?'

Reluctantly Jinda nodded. 'We were washing the buffalo,' she said.

'Well, when I'm not talking like a textbook, I don't know how to talk at all. Like just now,' he shook his head and stared out of the window. 'I didn't know how to ask you, just now, to stay on with me; in Bangkok.'

'You mean, you want me to stay?' Jinda asked.

Ned nodded, then shook his head. 'It's not that simple. For myself, I want you to stay. But for your own good, I thought maybe you'd be better off back in Maekung. All those rumours about the military staging a takeover . . . it's going to be dangerous here. I want you to be with me. But what can I offer you . . . promise you?'

Jinda listened in wonder, as the awkward, disjointed phrases tumbled out, so different from Ned's usual smooth flow of words. 'You don't have to promise me anything,' she said softly. 'I'd stay on if I just knew you wanted me to.'

'No, Jinda. I want something better for us. I want us to be able to build a life together. A home together.'

'Why . . . why can't we?' Jinda asked. There was a wisp of hair over his forehead, and she longed to brush it gently back, but she didn't dare.

'Because everything is so uncertain. If there is violence at the rally tomorrow, I could be arrested, or shot, or forced to go into hiding. And where would that leave you? I can't expect you to wait for me . . .'

'You can't expect it,' Jinda said, 'but I will. I will wait for you, Ned, no matter what happens tomorrow.'

Ned reached out and put his hands on her shoulders. 'Jinda, Jinda,' he said, shaking her gently, 'do you really mean that?'

She smiled. 'I never say anything I don't mean,' she said.

Ned's hands slid down her back, and he pulled her to him. Holding her tight against him, he rested his cheek on her hair. 'Jinda,' he murmured over and over again, 'Jinda, how can I say . . . how can I say this?'

She lifted her face and looked at him. 'It's easy,' she said quietly. 'I love you. Try it.'

Ned took a deep breath. 'I love you,' he said. Then he broke out into a broad smile. 'That was harder than any speech I ever made!' he said.

Jinda smiled back, then put her face next to his chest again. His shirt was cool and smooth, and Jinda could feel the slight rise and fall of his chest as he held her close.

A night breeze blew through the open window, and the wooden shutters creaked on their rusty hinges. A headlight from some passing car pierced the darkness, then was gone. It was very quiet, and peaceful. How wonderful everything is, Jinda thought. Father will be freed soon, and then Ned and I will build a home together. After tomorrow.

Chapter 12

Never in her life had Jinda seen so many people. From where she stood on the speakers' platform, she could see the crowd stretching from the Thammasart University gates clear across Pramane Square to the Temple of the Emerald Buddha. Like the incoming tide of a stormy sea, wave after wave of people streamed in, their heads bobbing underneath waxy green umbrellas. Easily forty thousand, Ned had estimated earlier that morning, maybe fifty. And more were still pouring in.

In the distance, Jinda saw a thick cordon of armed anti-riot police, their khaki uniforms looking incongruous under the striped canopies of the fruit vendors surrounding the square. Closer by were tight groups of street toughs, some of them displaying tattoos on their bare arms, others strutting around with hunting knives or ropes tucked in their belts. Were these, Jinda wondered uneasily, the gangsters reportedly recruited by the army to disrupt the rally?

Ned came up behind her, and seemed to sense her nervousness. 'Don't worry about your speech. You'll do fine,' he said softly. He pointed out a little boy to Jinda. He was, Ned explained, a shoe-shine boy who attended every rally held in the Pramane Grounds, posting himself on an empty carton to mimic the speakers. The little boy was standing on one now, waving his arms about theatrically, shouting at the top of his lungs.

'Imperialist dogs! Democracy brothers and sisters! Dictatorship feudalist!' He yelled enthusiastically. He made no sense, but the stream of political catchwords tumbled so fluently from him that he drew an amused audience, many of whom applauded and cheered him on.

'If he can do it, so can you,' Ned teased Jinda. 'With a bit of practice, you'll both be great orators . . .' Ned was interrupted by a sudden commotion nearby. Several street toughs had rammed their way through the audience surrounding the little boy, and were yelling at him to stop.

The boy, who looked a little like Pinit with his wide alert eyes, ignored them, and only raised his voice. Without warning, one of the men hurled a rock at him, hitting him on the forehead. Howling, the blood dribbling down his cheeks, the little boy jumped off his box and tried to run away. Another member of the street gang grabbed his collar and started pummelling him. Some bystanders tried to pull the boy away, but most stood there, immobilized.

In seconds switchblades had snapped open, pointed at those who were helping the boy. Immediately everyone stepped back, and the boy was left defenceless. He shielded his face with his arms as the men slapped and kicked him. One of them punched him so hard that he was spun reeling into the crowd. They let him go then, laughing as he crawled away.

Jinda watched horrified. The whole incident had taken less than a minute, and already the hoodlums were melting into the crowd. A few students started in pursuit of them, but Ned called them back. 'We've got to keep the rally peaceful,' he shouted. 'Don't give them any excuse for violence. There's too many of them out there.'

Only then did Jinda notice the number of anti-riot police

had increased so drastically that they now stood three or four deep around the entire square. Swarms of police had also gathered at the crossroads, cutting off the square from the Chao Praya river. Uneasy in the presence of so many more policemen than usual, the crowd was subdued under the bright flags and banners.

Dark grey storm clouds massed overhead, blocking out the morning sun. There was no hint of a breeze, and the air hung still and heavy. Jinda was oppressed with a sense of dread. She knew that this rally was crucial to the fight for lower land-rents, that her father's freedom, perhaps even his life, depended on it. Yet, Jinda felt now that she should have called the whole thing off if she could. Something was about to happen, she felt, something terrible.

But Ned had already walked to the podium, and was starting to talk. His voice boomed out, amplified by dozens of loudspeakers across the square, but Jinda was too nervous to concentrate. Partly it was because she knew she was to speak next, but it was also because of this awful foreboding. she felt as if she couldn't breathe.

The uneasiness had been building up in her over the past week, as she sensed the mood of the city turning ugly.

It had started with the military radio station. Every day, it broadcast 'reports' that student leaders had crossed the border into Laos and Vietnam to make contact with the Communist Party there. A sensationalist newspaper even printed a story about students who roamed the countryside killing farmers and siphoning their blood to build a blood bank for Communist guerrillas. Several military officers gave speeches claiming that the students were planning to topple the Thai king and queen, massacre all landowners, and plunge the country into chaos. The army would be needed, they claimed, to crush this plot. In the midst of these vicious reports, the dictator of the previous regime had quietly arrived by private plane, ending his exile from Thailand. Despite a law forbidding his entry into the country, he was living in an army base, biding his time.

Jinda thought through all this, and her sense of dread deepened.

Her heart skipped a beat now, as she suddenly noticed a convoy of trucks slowly drive in at the far end of the square, and more soldiers, shouldering rifles, leapt out of each one. Behind them a line of white ambulances waited, doors already swung open.

What's going on? Jinda thought in panic. She looked at Ned, and saw that he was holding his hand out to her. With a start, she realised that he was introducing her to the crowd, inviting her to the podium. It was her turn to speak.

Unsteadily, she walked towards him. A roar of applause greeted her. She waited for it to die down before she began her speech.

'I am Jinda, daughter of Inthorn Sriboonrueng,' she began. To her surprise she found that her voice was steady. It reverberated from all corners of the square, and even from the wall of the temple. She felt awed that her voice could reach so far.

She had practised the speech so many times that it had become automatic, and it flowed effortlessly from her now, separate and yet a part of herself. Like a kite with a lovely long tail, tugging its way upwards as she held the string, her words flew up. It was an exhilarating feeling, and Jinda's voice grew stronger with it. 'My father has farmed all his life,' she said, 'and yet he has never had enough to eat. Why?' She paused, and in that brief silence she felt that maybe, just maybe, she could help change a bit of Thailand after all. 'Because he's had to pay half his harvest to the landlord. Year after year. Flood or drought.'

'Commie bitch!' A shrill voice pierced the air.

Startled, Jinda stopped. Who had shouted that? Why?

'Keep talking!' Ned hissed.

Taking a deep breath, Jinda continued. Before she could

finish her next sentence, another obscenity was flung at her. Shaken, she tried to go on.

Suddenly, a heavy object sailed towards her, landing where the shoeshine boy had been. There was a loud explosion, and bits of dirt and glass shattered out. In the rolls of smoke which poured forth, people screamed, and started to run.

Ned grabbed the microphone from Jinda, and urged the crowd to be calm. 'Nothing serious has happened,' he announced. 'A small homemade bomb has just been tossed at us. This has happened in previous rallies. It hasn't hurt anyone. Do not panic, I repeat, do not panic. The speech will continue.'

But something was happening on the far side of the square. The soldiers in their olive green fatigues had fanned out in front of the ambulances, and were advancing towards the centre, pushing the crowd forward.

There was another explosion. It landed further away, but the bomb was deafening — and devastating. As the smoke cleared, Jinda stared, stunned. At least five students sprawled motionless on the grass.

Ned grabbed the microphone and tried to calm the crowd down again, but there was no curbing the panic that swept through them now. Screaming, flailing at each other, people tried to claw their way out of the square.

A few students, some of whom Jinda recognised as Kamol's friends, drew out their pistols and brandished them, then fired wildly into the air.

'No!' Ned shouted at the students. 'Stop! Don't shoot back!'

But they continued to shoot in the air, and that was when the distant gunfire started. At first Jinda did not know what it was, this sharp staccato rattle. Then she saw students dropping to their knees, in crumpled heaps, and she understood.

'Dear Lord,' Ned whispered. 'It's happening.' He bowed his head, and for a second just stood there, motionless.

'Ned!' Jinda clutched at his hand.

He snapped back into control then, and said, 'It's no use, Jinda. It's over. Run, quick. Along the river.'

'And you?'

He shook her off. 'Go!' he shouted. 'And if my house isn't safe, go to The Headquarters. Hurry!' He pushed her off the platform, and she landed clumsily on the grass below. 'Run, Jinda!' he shouted.

Jinda ran, but once she looked back, and saw Ned standing on the platform, alone, urging the people through his microphone to help the wounded. Then a knot of men climbed up onto the platform and struck him. He fell, and Jinda couldn't see him anymore.

She ran on, forcing her way through a heaving wall of bodies, their sweat and panic rubbing off her bare arms. Their voices were shrill with fear, and in their eyes was a blind terror.

Someone pushed past her. In his hand was a stick pierced through with a long nail. The nail scraped against her cheek and drew blood. Jinda suppressed her scream and ran on.

She reached a small clearing, next to the Thammasart gates. There was a big tamarind tree there, where she and Ned had often met after his classes. From the lowest branch of the tree now there swung a body. Jinda screamed. His face was hidden in the shadows, but his throat was rubbed raw by the thick rope, and his wrists were stumps, dripping red. A group of men ringed the tree, jeering. The man with the nailed stick shoved past them, and stood under the swinging body. As he lifted the stick and struck the dangling legs, the body swung towards Jinda.

Jinda couldn't watch. She fought her way past the crowd. Let me pass, she cried, I can't breathe, let me through. Face after face swam at her, but nobody looked at her.

At the gate of the university a huge bonfire had been lit,

tongues of flame licking at the sky. She stumbled past, trying not to look, yet looking. Splayed out, under the flames, was an outstretched arm, palm upturned, fingertips charred. A pair of broken glasses glinted there, reflecting the fire. My God, they're burning us, she thought. Beneath the flames, a tangle of feet jutted out, some barefoot, some with sandals still dangling. Scorched. Someone tossed a rubber tyre on top, and the flames danced higher than ever.

A knot of people approached the fire, dragging a body which left a trail of blood on the matted grass. Someone grabbed it by its legs, and someone else grabbed its arms. Together, heaving, they tossed it on top of the bonfire. Jinda ran.

Further on, other people were heaving a young girl into an ambulance, her white blouse flowered with splotches of bright red. The ambulance, Jinda saw, was already crammed full of bodies.

There was a gap in the cordon of khaki uniforms. She dashed through it, past the soldiers, past the crowds, past the sickening smell of burnt rubber and burnt flesh.

Someone screamed. She didn't stop. Keep running, faster. Faster. Further away now the rattle of machineguns, softer the occasional grenade.

It was less smoky here. The shady street was almost deserted. A few fruit-vendors squatted by their wagons, scared and bewildered. They stared at her.

It's the blood on my cheek, Jinda thought, and wiped at it with a shaky hand. She tripped just then on a basket, spilling out bunches of white lotus blossoms. An old woman stumbled out and cursed at her, her mouth streaming riveluts of crimson through the toothless gaps. Spitting blood?

Betel-nut, Jinda realised. A wrinkled old peddler with red betel-nut juice dribbling from her teeth-rotten mouth.

Jinda ran on. Down the pavement, against a red traffic light which no one heeded, and finally across the green metal bridge across the river to Ned's house. Underneath the bridge, muddy waters swirled past. Then Jinda saw it: a body draped in a white shirt, with a billowing black skirt. The face was hidden behind a curtain of tangled hair. Jinda did not look down into the river again.

On the other side of the river was a vegetable market. Its wooden stalls were deserted now, pyramids of eggplants and cabbages left abandoned.

Jinda ran past the market, and down the narrow alley, Ned's home, she thought, it would be safe there.

But it was not. A jeep was parked outside, and men in tight khaki pants were prowling through the house. She watched as they tore down the posters of Maotsetung and Che Guevara, and gathered up armfuls of Ned's notebooks and folders. One of the saw them tape recorder, and switched it on.

Jinda heard her own voice, clear and plaintive, singing Ned's favourite song. 'If I could be born a bird, with wings to fly, far far away,' Jinda heard herself sing, in a voice and in a time which now seemed incredibly remote, 'I'd ask to be a white dove, to lead my people to freedom.'

Jinda forced herself to walk on, slowly and steadily. Don't run, she told herself, don't panic and don't look back.

Behind her the song continued, and she strained to listen as she walked out of the alleyway and onto the main road again.

The tanks were out. A line of them rumbled down the road, barrels jutting from grey turrets like huge maimed insects groping with one antenna. She watched them roll past.

Where could she go? Where was it safe?

Headquarters, Jinda thought, that's what Ned said.

Jinda started to walk there. She blocked out all sounds and all thoughts, and started counting her steps. One two three four, she counted, all the way down the street. Vaguely she was aware of the faint gunfire behind her, but she kept walking, and counting. At six hundred and forty eight, she rounded the corner into the small alley leading to The Headquarters, and stopped.

An unreal sense of quiet, as sudden as the clarity of morning sunlight dispelling some nightmare, swept over her. A clothesline strung with baby clothes and sarongs flapped in the breeze. A group of boys were spinning tops on the cracked pavement.

The narrow, two-storied building of The Headquarters stood in front of her, its front door ajar. She slipped through it.

The neon lights were on, pale and cool. It was eerily quiet. The rows of desks were abandoned, the typewriters squatting mutely on them. In the far corner of the room, a group of people were huddled around a radio. They looked up as Jinda walked towards them. Kamol's face, tense and haggard, stared up at her.

Over the radio, a metallic voice announced that the Communist plot to take over Bangkok had been crushed. 'The situation is now under control. The Communists have been defeated. The army has saved Thailand from the clutches of the Communists. Curfew is at six. Repeat, the curfew is now . . .' Jinda walked over to the radio and flicked the switch.

In the sudden silence, her breathing sounded desperately loud.

'Well?' someone asked. It was Kamol, whom Ned had specifically asked to man The Headquarters today, so that he could not lead a violent retaliation against the opposition. Kamol looked haggard, his eyes ringed with circles so dark they looked like craters in his face.

Well what? She stared at Kamol.

'Say something,' Kamol snapped.

What is there to say? I can't say anything.

'The reports of shooting, and bombs . . . is it really happening?'

Jinda nodded. There was a potted orchid hanging by a window, and she stared at it. 'Shooting, and bombs,' she echoed. A sliver of sunlight pierced through one petal of the orchid blossom, making it look translucent. 'And people dying.'

Suddenly she was shaking, her teeth chattering. It's so cold in here, she thought. That's why I'm shaking.

Someone gripped her shoulders and guided her to a chair. She sank into it, sitting on her hands. She could feel their icy coldness on the back of her thighs.

'Here, drink this,' someone said, pushing a bowl of hot soup towards her.

Jinda shook her head. 'People dying,' she stammered, clenching her teeth to keep them from chattering. A piece of pork, edged with fat, floated on top. Dead meat.

The retch started deep in her stomach and forced its way up. She grabbed for the bowl and threw up into it. The sour smell of vomit permeated the room.

'Did Ned say what to do now?' Kamol asked, with a tinge of impatience. 'Can we join him out there?'

Jinda put her hands over her face. I don't know, she shook her head. I don't know anything anymore.

'Maybe we should pack up The Headquarters and leave right away,' another voice said.

'Yes, if it's that bad, the police are bound to raid this place any minute . . .'

Suddenly the door opened. They all jumped. Kamol was the first to recover, his laugh filled with relief. A frail young woman had walked in: Sri.

She was breathing heavily, her cheeks flushed, her short hair dishevelled. Yet she seemed calm. She walked over to Kamol and handed him a small bag.

'A roll of film,' she said. 'Pannada took it, wanted it developed right away and have the prints sent out to the foreign news agencies.'

'You saw Pannada at the rally just now?'

'I was treating her stomach wounds.'

'Where is she now?'

'Dead,' Sri said. 'Please develop that film now. She wanted it done immediately.'

Quietly Kamol took the roll of film, and went into the adjoining darkroom.

'Remember to burn what you can't pack. I'm going back to Thammasart. They're short of doctors there.' Sri headed for the door.

'Wait,' one of the students called to her. 'Sri, tell us. Is it really over now?'

Sri stared at the floor for a long moment. 'Only the patching up,' she said finally. 'God help me, only the patching up.'

She was halfway through the door when she stopped short, and pulled an envelope out of her shirt pocket. 'I almost forgot,' she said, 'somebody give this to Jinda if she comes in.'

Jinda felt a sudden cramp in her stomach. 'I'm here.' Jinda said quietly.

Sri swivelled around and stared at her. 'Oh no,' she whispered. 'I . . . I was h . . . hoping . . .' She was terribly pale, and seemed on the verge of collapse.

'You were hoping you wouldn't have to give it to me yourself?' Jinda asked. She reached out for the letter in Sri's hand.

Stumbling, Sri walked to Jinda, and knelt down on the floor in front of Jinda's desk. She opened her mouth to say something, but no words came. Silently she held out a white envelope towards Jinda. Her hand was shaking so badly she had to steady it with her other hand. 'It came yesterday,' she said at last, her voice barely audible.

Even before she had seen Pinit's rounded handwriting, Jinda knew what it was. She unfolded the letter. The single sheet of notepaper had been neatly ruled in pencil, and Pinit's writing was neat and clear.

'Please tell my sister Jinda,' it began, 'that our father died yesterday, on the Eighth day of the Fifth Lunar Month, in the year of our Buddha 2519. They brought his body back from the prison today. He looks a lot thinner. They said he died from a high fever. Grandmother says the funeral should be as soon as possible but we will try and wait for you. There are only the two of us at home now, Grandmother and me, and Grandmother cries a lot. Please tell Jinda to come home quickly. Sincerely, Pinit Scriboonrueng.'

The room was very still. It seemed as if no one was even breathing.

The silence was broken by the sound of sobbing, choked and muffled. For a moment Jinda thought she was crying, but then realised that Kamol had just come out of the darkroom. In his hands he held a sheet of contact prints, still shiny and wet.

Wordlessly, he placed it on the desk where Jinda and Sri were, and where the others now gathered to look.

The first photograph was of a young girl, her T-shirt pulled over her face, a knife thrust between small scarred breasts. The second was of the bonfire in front of the Thammasart gates — Jinda noted that the glasses by the fire were crushed now. The third was of rows upon rows of students made to kneel on the grass, under the raised guns of M-16s. There were others, many others.

The last one was of the tamarind tree in the clearing. In the photograph, the slim body had swung round so that the face was now out of the shadows. Jinda saw that it was the little shoe-shine boy.

The blood had made jagged streaks of black on his forehead and a line of it ran down the corner of his mouth. His neck was ringed with welts, and a coarse rope pulled taut the skin under his jaw. But his stumps of wrists had stopped dripping.

Kamol was crying openly now, his broad shoulders shaking, his thick lips twisted into a pathetic grimace. Jinda wished she could cry like that.

She got up and walked to the window where the potted orchid hung. In the street below boys were still spinning tops, and the noodle vendor was still slicing scallions. Everything has changed, Jinda thought, and yet nothing has changed. She reached for the single orchid in bloom above her, and twisted it off its stem. Petal by delicate lavender petal, she crushed it between her fingers and dropped it onto the street below.

Chapter 13

The valley unfurled below her in hues of green and gold. Encircling it was the range of mountains, its ridge tracing the arc of the sun's climb from east to west. Even as Jinda watched, the sun was slipping below the ridge, and soon the domed twilight sky would flatten into darkness.

Standing on the hilltop, she could see Maekung below her, its cluster of thatched roofs tucked among groves of banana trees. Radiating out from the village, with the imperfect symmetry of a spiderweb, were the rice fields, still brown, still dry, still barren. Only the stubble of the last harvest was left, its few stray stalks long since gleaned bare by scavenging crows.

Had she really been gone? Perhaps time, like water in the deepest section of a river, flowed more slowly here. Perhaps she had never gone to Bangkok at all, never lived there and worked there and witnessed the horror of the student massacre there. Perhaps she had just been out in the forest gathering mushrooms all day, and was now returning home in the twilight?

She slung her pack over her shoulder, and started walking downhill. The path was loosely pebbled, a gentle slope twisting between rock outcrops. She passed the slab of granite where she had often watched butterflies hatching from their cocoons; then the knotted vine dangling from the raintree where Dao had taught her to swing.

Jinda quickened her steps on the last stretch home. She could imagine her little brother drawing water from the well, and the look of surprise on his face as she called to him. Then her grandmother would hobble out of the kitchen and peer at her, and Father would run down the steps, tall and strong and laughing, arms outstretched.

Suddenly Jinda's knapsack felt unbearably heavy. No, she thought, not Father. Father would never run down the steps to greet her again. She slowed down: what was the point of hurrying home anyway?

Dragging her feet, Jinda slowly walked down into the valley. When she reached the village, she found it strangely quiet. A skinny rooster scratched at the dirt. It was the only sign of life.

She walked down the path through the village. The shutters of most of the houses were closed, and the yards were deserted. It was as if the whole village was in mourning.

And then she was standing in front of her own house. It looked smaller, somehow shrunken and sagging. A thick layer of dust coated the wooden boards, and limp morning glory vines draped the fence.

No one was in the yard, but a steady hum of voices, as subdued as the first evening cicadas, drifted out of the house. Jinda slipped past the gate and stood under the mango tree, where she could watch without being seen.

Several lamps were lit, and knots of people huddled on the verandah — mostly women. Heads bent, their fingers fluttered in the light like nervous moths. They were wrapping, folding, tying things that Jinda could not see.

An old woman glided down the steps and padded over to Nai Wan's yard, where she sliced off a few pieces of banana leaves. Back on the verandah, she handed the leaves out to the various groups. And only then did Jinda understand.

Her father's funeral would be tomorrow. So soon. The women had gathered to make the final preparations, to wrap the fermented tea leaves with squares of oiled banana leaf; to fold paper boats and arrange the flowers and candles in each one; to tie little bouquets of pine kindling and incense.

And her father? Where was he? Lying alone in some dim corner of the temple, shut up in a box? It did not seem possible. Jinda wanted to go to her grandmother and ask, but she could not face the throng of people on the verandah.

And so she remained hidden in the shadows of the mango tree. Soon it was dark, and many of the women rose to leave. A few bustled about, sweeping and picking up things, but even they murmured goodbyes and slipped off. Finally her grandmother was left alone.

The old woman went into the house, but soon she reappeared again, carrying a stack of clothes in her arms. Her every movement slow and deliberate, she sat down on a mat, and laid the clothes next to her. She pulled the kerosene lamp to one side, and her sewing basket to the other. Carefully, she threaded a needle and picked up the first shirt from the pile. It was Inthorn's.

Hunched over the lamplight, she sewed neat little elbow patches on it, then started to mend the frayed collar. Though a night breeze was blowing, she wore nothing but a thin undershirt, its two straps occasionally slipping down over her bare shoulders. Her grey hair was tied in a knot at the back of her head, leaving the nape of her neck delicately exposed as she bent over her sewing.

Having mended that shirt, she folded it and reached for another. It was much like the first one, bleached by years in the sun and countless scrubbing on the stones of the river bed. This one, too, she carefully mended.

One by one she went through the little pile, never once pausing or looking up.

The last piece of clothing was Inthorn's jacket. She had made it for him herself, and it had worn well. She held it up in the light now, examining it closely. Only a button from its sleeve was missing. She sewed on a new button, taking great care to align it with its buttonhole. When that was done, she started folding it up then stopped. She closed her eyes, and held the jacket to her cheek, rocking it against her gently, to and fro, to and fro. Her thin shoulders shook, but the thick coat muffled the sound of her weeping.

Grief has many tastes, but for Jinda, that night tasted of mango. Watching her grandmother hug that coat, Jinda crammed handfuls of mango leaves into her mouth so that her grief, too, might be silent. The crushed leaves tasted of unripe mangoes, bitter, puckering her mouth. But she chewed them silently until her grandmother's shoulders had stilled. When the old lady lifted her face again, it was twisted into furrows of pain. Breathing deeply, she carefully folded the jacket and laid it over the pile of mended clothes. When she had finished, her face was calm and composed again.

Only then did Jinda step out of the shadows of the mango tree and call to her grandmother.

The funeral took place the next afternoon. That morning, as they watched skilled carpenters build a miniature temple of plywood and tinsel for the coffin, Pinit told Jinda how their father's hand, wounded on the day of his arrest, had apparently festered in prison. Jinda listened impassively. It did not matter anymore. Her father was shut into a tight wooden coffin now, and details about how he had died seemed unimportant to Jinda.

When the miniature temple was finished, it was set over the coffin, which was then hoisted onto a huge cart laden with flower wreaths and candles. A thick cord was tied to one end of the cart, for the monks and mourners to pull it to the cremation grounds.

Jinda held the cord with both hands now, but she did not help pull. In a daze, she walked near the head of the cortege, next to Pinit and their grandmother. She was only vaguely aware of the people around her.

So that she would not have to look at the coffin, Jinda stared at a single piece of tinsel which had unwound from the paper temple mounted over the coffin. It fluttered sinuously in the breeze, a strip of silver against the blue sky. Then it caught on a branch, and was snapped off. It dangled there, suddenly limp. Lifeless, Jinda looked away.

Most of the villagers had joined the procession, the men solemn in their starched workshirts, the women severe in their black blouses and black sarongs. Nai Wan, Nai Tong, Bamrung and several of her father's closest friends were up ahead, stumbling over the rocks on the path as they strained to pull the cart along. Their shirts were already soaked through. Behind them, their wives walked with lowered eyes, holding squares of banana leaves to shield their faces from the afternoon sun.

The cremation grounds were in sight now, dappled with shade under the canopy of heart-shaped Bodhi leaves. Knots of villagers broke off from the procession and hurried towards the shade, dropping down to squat on the gnarled roots of the Bodhi trees. Jinda remained in the sun, next to the coffin, until the cart had been dragged into the shady grove.

It was cool and restful under the Bodhi trees, and a light breeze brushed against her bare arms, drying the prickles of sweat there. She thought of her father boxed up in his narrow coffin, and for a moment wished he could be brought out to enjoy the breeze.

The musicians started up, three men swathed in loose black clothes, each with a white sash around his waist. One played a lute, another a wooden xylophone strapped around his neck, while the third stroked a steady rhythm from a long, light drum. The music, at first a mournful metallic wail, grew louder and shriller until the hot air throbbed with it.

Then the monks started chanting, their sonorous voices competing against the funeral music. Once Jinda would have listened to the stream of incomprehensible Pali words with respect, but today their chanting seemed hollow. The monks were supposed to help send her Father's soul to a better life in his next incarnation, but what, Jinda thought bitterly, what had they ever done for him in this life? Had any of them ever chanted for his release from his prison cell?

Like polished gourds, their smooth-shaven heads glistened in the sun. One of the novices was scratching his armpit, and another kept swatting at the flies around his ankles.

Their chanting finally over, the monks sprinkled blessed water into the little banana-leaf boats at each corner of the coffin. Then they withdrew, the hems of their orange robes fluttering around them as they walked.

A group of farmers came forward, hoisted the coffin off the cart and set it on the ground next to the crematorium. The wooden lid was then lifted off.

It was time for the last farewell.

Jinda's grandmother walked towards the open coffin. Her face was impassive. Only her stooped shoulders betrayed any sign of strain. When she reached the coffin, she looked into it for a long moment. Then she laid down her cane, and dripped her hands into a bowl of blessed water. Carefully, she sprinkled the water into the coffin.

She called to Pinit to join her. Jinda watched her little brother as he, too, stood looking into the coffin. His young face was delicate and open, and in his eyes was a bewildered pain. He did not cry, yet there was a darkness over him, as if a passing raincloud had cast a patch of shadow only on him.

The boy took a freshly opened coconut and prayed. As Inthorn's son, it was his role to cleanse his father with coconut water, because the clear sweet liquid was pure, untouched by human hand. Gravely he poured a trickle of the coconut water into the coffin. When he was finished, he looked up and searched the crowd for Jinda, his round eyes pleading.

Jinda stepped forward then, and took the coconut from him. Her hand trembled so badly that a few drops of liquid splashed out. She could not bear to look into the coffin.

This isn't him, she told herself fiercely. This is not my father anymore. Father has gone, Father is free. This is just the shell, like an empty cocoon after the moth has flown away.

Jinda steeled herself and looked.

And there was really nothing of her father there, nothing but a shell. The hair was greyer, the face more gaunt, the bare feet bony and barely familiar, the toenails edged with dirt that might have come from these fields here. The hand — here Jinda had to turn away. Swollen and putrid, the hand was raw where the flesh had burst through the brown skin.

I should have washed it more carefully that day, Jinda thought with fierce regret. It is too late now, but I will wash it one more time. Her hands steady now, she poured the water from the coconut down onto the mangled hand. A few drops of liquid gathered in a little sparkling pool in his cupped hand.

The lid to the coffin was put back on, and the coffin itself hoisted up to the low brick walls of the crematorium. Jinda helped her grandmother back to the shade, and waited.

Soon the oldest monk lit the white thread leading to the funeral pyre. The thread burned its way past several branches and finally sparked off the oil-soaked kindling beneath the coffin. Within seconds flames leapt up, reaching high into the sky.

The flimsy temple mounted on top of the coffin caught fire, its delicate paper columns sending tendrils of flame up to the tinsel roof and eaves. As Jinda watched, the graceful roof collapsed in a burst of flames onto the coffon, and only a charred skeleton of the pretty temple was left standing, lopsided and crooked.

Among the crowd, trays of funeral bouquets, delicate fans woven of pine kindling and incense, was passed around. Jinda took one, and was swept up by the crowd towards the pyre. The heat of the flames pressed against Jinda's face and bare arms as she approached.

The coffin was blackened and smoking now. Soon it too, would catch fire.

People were tossing their pine bouquets into the fire, then backing away. Jinda climbed the three steps towards the open fire, and looked into it. The fire was burning so fiercely now that she couldn't even see the coffin anymore, engulfed now by huge dancing tongues of flames.

Jinda threw her pine bouquet into the fire, and it too, was swallowed up. She turned away, her eyes stinging.

The ceremony over, most of the villagers drifted off in little groups, out of the Bodhi grove to the dry heat of the afternoon sun. Some of them broke into light chatter as they walked single-file along the mud dykes of the fields. The monks too, scattered, their bright orange robes like sparks of ember blown from the funeral fire across the bleached countryside.

Around the pyre only a few farmers remained, squatting in a small circle. Pinit was among them, huddled against Nai Wan. But nobody laughed or talked. They did not roll dice over a bottle of rice whisky, nor talk with gruff tenderness about the dead. This was, after all, not an ordinary funeral, of some old man who had lived out his life fully and died peacefully.

Father died young, Jinda thought, and he died painfully. Why? Leaning against a big Bodhi tree, she pressed her cheeks onto its rough bark. What had it all been for? A few more bushels of rice? Surely her father's death wasn't worth that. But Ned would claim that he had given his life for a cause, an idea. Justice, equality, democracy, what were those big words Ned always used? They seemed so remote now.

Jinda stared at the fire. A few flakes of ash, skimmed up by a passing breeze, scattered into the air. One flake blew past her, and she reached out and caught it. When she opened her hand, there was only a faint smudge of grey on her palm.

From flesh to ash, from blood to smoke — her father had died for an idea. With bitter sorrow, Jinda realised she didn't even understand what the idea was about.

In the dry hot days after the funeral, Jinda never returned to the Bodhi grove where her father's ashes laid. Even when her family went to collect his bones the day after the cremation, Jinda stayed away. She hated the thought of her father reduced to the crumbs of charred bone now enshrined in that little urn on the family altar. Jinda did not want to go back there, and she did not want to brood over her father's death. He was gone, and no amount of mourning would ever bring him back. She was the head of the household now, and it was up to her to plan for the family's future.

Relentlessly, Jinda forced herself to work. In the month that she had been away in Bangkok, their house had fallen into even worse disrepair than before. Many of the planks on the verandah had warped or rotted through. The roof needed rethatching, and the bamboo fencing was sagging under the weight of untrimmed vines. And soon, if the rains came, the seedbed for the rice would have to be prepared, and the fields ploughed.

Jinda set about each task with fierce energy. Sometimes, caught up with her work, she would look up and notice others sitting idly about, and glare at them accusingly. Once, as she was drawing water from the well, she saw Pinit nearby, his chin cupped in his hands.

'What're you doing?' she demanded.

'Nothing.'

'Well, why don't you do something?'

'Like what?' Pinit asked timidly.

Jinda poured the well-water into a jar, splashing much of it on the dry sand. 'I don't care!' she cried. 'Just don't sit there!' Flinging the bucket down, Jinda stalked off.

'Why is she always like that now?' she heard Pinit asked their grandmother as she walked out the yard. 'Is she angry?'

'No, child,' the old woman said. 'She's just determined not to be sad.'

That night, Jinda did not sleep well. Usually, she would have driven herself close to exhaustion at the end of each day, so that sleep would come with the darkness. But her grandmother's words had disturbed her, and she could not sleep. Was she angry? Was the anger really just a shield against a deeper sorrow?

For at night, the memories came back, and with them the flood of bitterness. She laid awake, remembering the killings at the rally, reliving her visit to her father in prison. She imagined Ned in shackles, in prison too, or lying wounded among the dead at the rally. She worried about the future, whether the rains would come on time and whether she'd be able to find the help she would need to plough the fields before the rains.

If Dao were here, she thought bitterly, she could help. But no, her sister had apparently left Maekung and her family behind her forever. She never answered Pinit's letter telling her of Father's death, nor did she even bother to send a wreath or some money for Father's funeral. She's probably forgotten us all by now, living her life of luxury in town with 'Mr. Dusit.' Well, go ahead, Jinda told Dao fiercely, we don't need you. We don't even want you back.

The next day, with dark circles under her eyes, she felt too tired and miserable to do any of her chores. Instead, she climbed up to a spot on the hill which she used to think of as her retreat. It was peaceful there, and commanded a view of the valley on one side, and of the dusty road leading to it on the other. It was here, she remembered, that Dao had also come after her baby's cremation.

Jinda leaned back against the teak tree now, and stared at the landscape. The stretch of bone-dry fields was relieved only by clumps of tassled weeds whose feathery plumes swayed in the breeze. Here and there, a charred tree stump dotted the hillside.

Was May always as dry as this, Jinda wondered? Had her stay in Bangkok, short as it was, made her forget how bleak the countryside was in the hot season? No, it wasn't that. The years of drought had taken their toll, and it was much drier and hotter than she had ever remembered. Another year of drought, Jinda thought, and we're all dead.

Idly, she watched a black beetle attacking a small dragonfly on a teak leaf next to her. The dragonfly tried to crawl away, dragging its torn wing, but the beetle chased after it. Jinda picked up a pebble and crushed the beetle. There, she thought, at least I saved the dragonfly. She looked at it, twitching its torn wing weakly. With the same pebble, she reached out and crushed the dragonfly too.

She tossed the pebble away and watched it roll down the hill. Suddenly Jinda sat up straight.

A tiny figure in the distance was climbing up the hillside on the path. Ned? Could it be Ned at last? Wasn't that what she had really been waiting for, all along? Her heart was pounding.

Jinda squinted her eyes against the morning sun. The figure moved slowly, stopping to rest every few steps. Jinda saw now that it was a woman. Not Ned, she thought with a stab of disappointment. The woman paused in the shade of a teak sapling, and lifted up both arms to retie her hair. Jinda immediately recognised the familiar gesture.

'Dao,' she murmured.

Dao had almost reach the spot where Jinda was now. Her

clothes were soaked through with sweat, and her breathing came in laboured pants as she approached.

'So,' Jinda said coldly. 'You've come back. Welcome.'

Dao looked up, startled. 'Jinda!' she gasped, then broke out into a big smile

Jinda stared at her impassively. 'I'll help you with your bag,' she said, in the same cold tone.

Dao shook her head, and dragged herself up the last few feet towards the shade where Jinda was. She sank down on the log there, hugging her big bag to her. For a moment neither of them spoke. Dao gazed down the other side of the hill, at the valley and Maekung village. 'It's good to be home,' she said finally.

'You should've come earlier, then,' Jinda said. 'For Father's funeral.'

'Jinda, I tried, believe me. I tried.' Dao stole a look at her sister. 'When I got Pinit's letter, I wanted to leave right away, but Dusit . . .'

'Dusit?' Jinda echoed. 'Not "Mr. Dusit"?'

Dao saw the look in her sister's eyes, and glanced away. 'How can I explain?' she asked wearily. 'So much has happened.' She wiped her sweat-glazed face on a sleeve, and her hand fell listlessly on the log.

Jinda glanced at it. Strangely enough, her sister's hand was bare. Jinda had imagined Dao's fingers to be flashing with rings like the gold one she was wearing when she left. 'Where's your fancy ring?' she asked.

'I pawned it,' Dao said simply. 'I needed the money to buy the ticket home.'

'Why didn't you just get your "Mr. Dusit" to give you the money?' Jinda demanded, thinking of Dao's first few letters home, with their thick wads of money enclosed.

'He hasn't given me any for a long time,' Dao said quietly. 'Can't you tell?'

Jinda took a good look at her sister then. Her hair was streaked brown with dust, and her clothes were shabby and wrinkled. She looked more as if she was coming back after a hard day in the field, than as a pampered mistress of a wealthy man in town.

'All right, what happened?' Jinda asked, more gently now.

Dao spoke awkwardly, and the story as she told it was a broken, confused one. It seemed that Dusit had treated her royally at first, buying her expensive clothes and renting her an apartment with fancy furniture and running water even. He had also assured her that he would arrange for special permission for her to visit her father in prison. It was weeks before she actually saw Inthorn, but when she tried to pass him some medicine, it was again confiscated.

'Father looked much weaker than when we saw him, Jinda,' Dao said, her voice shaking. 'Thinner too. He could hardly walk, with those shackles dragging at him . . .'

'And after that?' Jinda said roughly. 'Didn't he get better at all?'

'I never saw him again,' Dao whispered, avoiding Jinda's eyes. 'I don't think Dusit ever really meant me to visit Father regularly. It was just one of the many false promises he made to get me to live with him.' She paused, and shook her head. 'I tried seeing Father again, of course. Dusit got angry whenever I brought the subject up, so I began asking around the prison and police stations myself. When Dusit found out I'd been trying to see Father behind his back, he got furious. He accused me of siding with the leftist students, and betraying him. So I never dared to try and see Father again.'

How can that man mean more to you than our father, Jinda wanted to scream. But she bit her lips, and stared at the sand.

'Later, I heard rumours that Father was very sick, but Dusit wouldn't tell me anything. We weren't even talking much anymore. Then Pinit's letter came, saying Father had died. When I told Dusit I wanted to come home for the funeral, he said that if I left, I needn't ever come back to him again. So I didn't come.'

'How could you?' Jinda burst out.

Dao put her hand on Jinda's arm imploringly. 'Don't hate me, sister. I know I was weak, and selfish. But Dusit meant so much to me then. I wanted so much to start a new family.'

Jinda stared at her. Wordlessly, Dao pushed the bag away from her, and put both hands over her stomach. Her sarong was stretched taut over a round bulge.

'You're pregnant,' Jinda said.

'I was sure Dusit would love me more, if I was bearing his child.'

'And he didn't,' Jinda said. It was not a question.

'Dusit likes his women young and slim,' Dao said drily. 'The ones he began to see after I became pregnant, were all younger and slimmer than I was.

'And then?'

Dao looked at her sister, and smiled. 'Do I have to spell it out for you?' she asked. 'He came and told me yesterday to clear out of the apartment. Someone else was moving in.' She paused, and tried to laugh. 'So I came home,' she said, simply.

Another mouth to feed, Jinda thought. And, judging by the size of Dao's belly, she would not be able to help with the heavy farm work either.

As if sensing Jinda's thoughts, Dao said, 'I can still work, you know. With the seeding, and repairing nets, and . . .'

'You'd eat a lot more than your work would be worth,' Jinda said coldly.

'Oh Jinda, have pity,' Dao murmured, hunched over her bulging stomach. 'If not for me, at least for this little unborn one. It hasn't done you any wrong. I want it to grow up in a real home, with family around to care for it. I want Granny to rock him to sleep, and Pinit to play with him, and you to help bathe him. Just as we used to bathe little Oi together, remember?'

Jinda nodded. She looked at the range of mountains brooding over their little valley, and the shadows they made across the rice fields. It was the same pattern as the afternoon little Oi was cremated. 'I remember,' she said quietly. She reached out and brushed the dust away from her sister's shoulders. 'Welcome home,' she added, and this time she meant it.

Chapter 14

Dao slipped back into village life so quietly it was as if she had never left. Her grandmother had refrained from asking her any questions, and simply welcomed her back with open arms, her lined face crinkled into a smile. And if some of the village women tended to gossip about Dao behind her back, a sharp look and a few caustic words from the old woman quickly silenced them. Within a few weeks, Dao was back with her old group of friends, drying chili peppers in the sun or mending clothes as she chatted the afternoon away.

And so the days slipped past, but still the rains did not come. By the end of June, like many of the other farming families, Jinda and Dao decided to start the seedbed anyway. If the rains did come, however late, at least they would be ready to transplant.

And so one evening Dao soaked two buckets of unhusked rice grains into a shallow basket of water, skimming off the empty grains that floated to the top. The remaining grains she poured into another basket lined with damp straw, and left them to germinate.

Meanwhile Jinda ploughed and harrowed a tiny plot in

the fields, careful to have the raised seedbed surrounded by a narrow ditch for irrigation.

When the rice seed sprouted tiny green shoots, the two sisters dragged the basket to the fields and broadcast the seeds, carefully tossing handfuls of the damp grains onto the seedbed. For days, they watched over the seedbed anxiously, waiting for the grain to take root and send the new shoots straight and strong above the mud.

The next step was the hardest. Ploughing the dry fields beyond the seedbeds was traditionally a man's work, and in their family it had always been their father who had done it. But now that Inthorn was dead, and their brother still too young, Jinda and Dao decided to do it themselves.

Together, the two sisters yoked the plough to the buffalo, and tried to plough. It was back-breaking work. The soil was still dry, and the buffalo too weak from hunger to pull with much strength. Jinda wiped the sweat streaming from her face now and leaned against the plough. The furrow in front of her was shallow and crooked, not at all like the deep straight lines her father had carved into the soil with such seeming ease. She had been ploughing the whole morning, yet she had barely finished half a small field. The buffalo had ignored her desperate cries of 'Right! To the right!' and had once even charged right across a bund into an adjoining field.

The main problem, of course, was the hard dry soil. The monsoon rains should have come long ago, and there had been a few scattered showers, but then — nothing.

Jinda looked up and scanned the sky anxiously. There were a few grey clumps of clouds massing behind a mountain ridge, but not nearly enough for the long heavy downpour they would need before the fields could be ploughed the second time, and then harrowed. Already the delicate seedlings were starting to shrivel in the seedbeds. What were they going to do if the rains didn't come?

Dao wiped the sweat from her face, and looked up at the sky too.

'Still no sign of rain?' she asked.

Jinda shook her head gloomily.

'It's hopeless, Jinda,' Dao said. 'We can't manage by ourselves. I'm no use to you like this,' she pointed ruefully at her bulging stomach. 'And besides, Granny says that the baby may come early if I keep straining. We've got to have some help, sister.'

'Sure,' Jinda said wearily. 'But who? Nai Wan and Lung Tong?' Both men had disappeared after Inthorn's funeral. Their wives would only say that they had left to find work in the city, an unconvincing lie since this was the busiest season of the year for farmers. Jinda and Dao both believed that they had joined the Communist rebel forces in the mountains nearby.

'Actually,' Dao said, 'I was thinking of asking Vichien to help.'

'No,' Jinda said.

'He's young and strong. He could plough two fields in the time you took to do one, and do it better too.'

'No.'

'He probably wouldn't even charge us for it.'

'I said no!'

'Why not?' Dao asked angrily.

'You know why.'

'I know that he likes you, has probably even asked you to marry him,' Dao said. 'What I don't know is why you won't accept him.'

Jinda did not answer.

'It's because of Ned, isn't it?' Dao asked. 'You're still waiting for him.'

'What if I am?' she said in a low voice.

Grasping her by the shoulders, Dao forced Jinda to look up. 'Listen, don't wait anymore, sister. He won't turn up. I waited for Ghan for months, hoping to show him our son. He never came. And then when I came back this time, pregnant, I waited for Dusit. That's right, I waited for Dusit

— even though I knew he wouldn't come. Well, not anymore. I'm not waiting for Ghan, or for Dusit, or any man.'

'Ned is different,' Jinda said.

Dao laughed. It was a brittle, sharp sound. 'Men,' she said flatly, 'are all the same.' She smiled thinly. 'At least you're not pregnant. And you've still got Vichien.' She shook Jinda's shoulders gently. 'Forget Ned, sister. Vichien is here. He can take father's place and help us work the land. We need him. Please!'

Jinda turned away. At the edge of the valley the mountains loomed, stark and brown. How many times had she gazed at them, watching for some sign of Ned, hoping to see him climbing down the trail again. But he had not come, and, Jinda thought bitterly, her sister was probably right: he would never come. Even if he wasn't in prison, even if he hadn't been killed at the rally that day, even if he was healthy and free, he had no reason to come back to Maekung.

She had written to Sri twice, asking for news of him. The first time, Sri had replied in a note so short and cautious it was almost curt. She was now working at a private hospital, she said, and did not keep in touch with any of her old friends anymore. She knew nothing of them, nor did she want to know anything. The second time Jinda had written, Sri hadn't replied.

Jinda looked at her sister now and took a deep breath. 'All right,' she said grimly. 'Go and ask Vichien.'

'Oh Jinda! I will!' Dao murmured. She looked immensely relieved.

'Just to help,' Jinda said. 'Ask Vichien, but just for his help. Don't you dare talk about marriage.'

Jinda leaned on her plough, and watched her sister waddling clumsily across the fields towards Vichien's house. She knew that asking for Vichien's help knowing full well his interest in her, amounted to accepting him. She would be obligated to him. She would be made to feel that she owed him something.

Jinda kicked at a clump of sod, but it was so hard that she stubbed her toe on it. Tears stung her eyes. If only Ned would come, she thought miserably.

She tugged at the plough, and was about to urge the buffalo to start, when she heard a rustling in the bamboo grove nearby. She thought she heard someone calling her, softly. Jinda turned around, but saw nothing. Again she heard her name, this time more urgently. 'Over here,' the voice whispered.

Frowning, Jinda left the plough, and walked to the bamboo grove. Someone was standing in the shadows, waiting for her. As she approached, the person stepped forward. Jinda gasped.

'Ned,' she said, her voice hoarse. She could hardly recognise him.

His face was gaunt, his clothes torn, and in his eyes was a hunted look. Around his neck hung an amulet on a string, a small statue of Buddha encased in plastic. His dark blue workshirt flapped open, its strips of cloth left untied. He had grown so thin that his shoulder blades jutted out sharply.

He glanced furtively behind Jinda. 'Nobody saw me?' he asked.

Jinda shook her head. She did not trust herself to speak.

'Please, could . . . could I have a drink of water?'

Gone too was the easy confidence in his voice when he had spoken around the campfire, or at the rally. His voice now was strained, as if he was used to whispering and was now making an effort to speak up. Standing there in the shadows of the bamboo grove, he seemed years older, and immeasurably sadder.

Jinda turned away. 'I'll get some water,' she said, not looking at him.

When she returned with a dipper of water, Ned was squatting on the ground, his arms hugging his knees. He accepted the dipper from Jinda with both hands, and lifted it to his mouth. He drank in gulps, the water trickling out of the sides of his mouth and staining his dust-streaked neck. Only when he had finished the last drop did he look up and thank Jinda. 'Come, sit down,' he said, patting the ground next to him.

Jinda remained standing. 'You've changed,' she said.

'It's been a rough few months,' he said, and tried to smile.

'Where have you been?'

'Moving around, seeing friends, thinking.' He paused, and stared at the ground. 'Hiding, mainly,' he said.

Jinda dropped down next to him then, and took his hand. Large and bony, with its big smooth knuckles, it seemed reassuringly familiar. 'I'm glad you're safe,' she said. 'I thought you might've been arrested, or imprisoned, or badly hurt . . .' An image of the little shoeshine boy hanging from the tamarind tree flashed across her mind. 'One boy, they cut his hands off and . . .'

Ned grasped her hand and pressed it. 'Hush,' he said. 'It's over now. Don't think about it.'

'I try not to, but it's hard sometimes,' Jinda said. 'Some nights I wake up crying, the dreams are so bad . . .'

'It's over,' Ned repeated. 'Those they've killed are dead. Those who've escaped are free. Like us, Jinda. It's up to us to start all over again.'

'Yes, start all over again,' Jinda said softly. She felt her heart lifting. They would start over again, and this time they would work together to build a home. The fields could be ploughed, and the rice seedlings transplanted once the rains came. How full her life would be, with Ned to share it with.

'He'll do it, Jinda!' Dao called across the fields. 'Jinda, where are you?'

Jinda sprang up and parted the bamboo leaves. Dao was making her way laboriously down a furrow towards the abandoned buffalo. Quickly Jinda walked out of the bamboo grove and hailed her sister.

Dao was in very good spirits. 'Did you hear me,' Jinda? I said Vichien will do it. He'll plough both our fields, and for free. Isn't that good of him?'

'We can pay him for it,' Jinda said.

'I told him that, but he said he wouldn't take our money.' Dao smiled broadly. 'What he meant was, your money, Jinda. He's really doing it for you.'

'I don't want that,' Jinda said.

'Oh sister, don't be so stubborn. Vichien's a good man. And he'd make you a good husband, you know he would. Don't waste your time waiting for that Ned. He won't come. I know more about men than you do, Jinda, and I'm telling you, that Ned is unreliable and he will never come.'

Ned stepped out of the bamboo grove then and faced Dao. 'I may be unreliable,' Ned said quietly, 'but at least I have come.'

Dao was so startled that she stumbled backwards. She tried to grab onto the plough for support, but her sudden motion only frightened the buffalo. The animal jerked away, yanking the wooden handle of the plough across Dao's legs. Uttering a sharp cry. Dao fell heavily over a row of rice stubble.

'Dao!' Jinda cried, rushing to help her up.

Dao's face was twisted with pain, and she clutched at her stomach, moaning softly. She looked dazed. 'I . . . I'm all wet,' she said.

Only then did Jinda notice a dark puddle soaking into the dry dust between Dao's legs.

'You're bleeding,' Jinda said, scared.

'It doesn't hurt,' Dao said, staring at the wet spot in the earth. 'It isn't blood. It's water. My water-bag must have broken.'

'Does that mean the baby might . . . the baby might . . .' Jinda stopped. She had no idea what the baby might do.

'The baby might come early,' Dao said tersely. Awkwardly, she tried to get up, but was too clumsy to manage.

'Help me lift her,' Jinda told Ned.

Together they half lifted, half pulled Dao upright. Her face was beaded with sweat, and strands of wet hair stuck to her cheeks.

'We'd better get you back home,' Jinda said. 'Granny will know what do to.'

'That's all right, I don't need your help,' Dao told Ned as he supported her elbow. Leaning just on Jinda, she started to walk, but again stumbled.

Without a word, Ned took hold of Dao's elbow again, and firmly helped guide her across the fields home.

Once home, their grandmother immediately took control of the situation. Giving a series of brisk orders, she had Dao settled comfortably on a padded mat near the window within minutes. Only when she had examined Dao, and decided that there was no serious injury, did she even acknowledge Ned's presence.

'Thought we'd be seeing you again,' she said laconically. 'Have you come back to hold of your meetings?'

'No, Grandmother,' Ned said, subdued. 'I doubt that I'll be holding any more meetings in Maekung.'

The old woman looked at him sharply. 'Why d'you come back, then?'

'To see Jinda,' Ned said.

The grandmother took one long look at Jinda and Ned standing together in the doorway, and nodded. 'Well, what're you two waiting for? Don't you think I can take care of Dao by myself? Go on, be off with you.'

Flashing her a grateful smile, Jinda led Ned down the stairs, and to the small path towards the river. The sun was setting, and they would not be likely to meet any villagers there.

'Where're we going?' Ned asked.

'Somewhere safe,' she said. Having sensed that Ned did not want to see anybody else, she had decided to take him to the big willow tree on the opposite bank of the river, behind whose spreading leaves she had once seen Dao and Dusit. How Dao would mock me, she thought, if she knew I was taking Ned to the same spot she had taken Dusit.

The space behind the trailing willow branches was cool and shady. Jinda found a mossy patch of ground, and sat down on it. Ned sat facing her, an arm's length away. For a moment they looked at each other in silence.

The Ned's stomach growled. It was a loud, rumbling sound, and Ned flushed, looking embarrassed. 'I . . . I've been too busy travelling today, to eat anything,' he explained lamely.

'I thought you might be,' she said. Before they had left the house, Jinda had packed her shoulder-bag with a little rice-basket and some fish-sauce, and she took these out now and offered them to Ned.

As wordlessly and quickly as he accepted the water, Ned stretched out both hands for the food now. Jinda watched him eat, and her heart ached.

Here is the man who told us we need never be hungry, she thought. We were used to being hungry, and never dared to want more. Until Ned came and urged us to fight for more — more land, more food, more "rights". And after all the fighting is over and done, we're still hungry. The only difference is that he's hungry too now, like the rest of us.

And yet, watching him finish every last grain of rice, Jinda felt no bitterness or contempt, nor even pity. He's really like one of us now, she thought. It'll be easy for him to settle down in Maekung, and together we can build a home here.

She realised suddenly that he was staring throughfully at her, and she laughed nervously. 'What're you thinking about?' she asked.

'About the last night we were together,' he said. 'About how we talked of building a life together after that rally.'

Jinda smiled. 'That's what I was thinking about too,' she said.

The words flowed from both of them easily after that. Jinda told him of her father's death, and the funeral, and Dao's return home, and Ned talked of the two months he had spent, living like a fugitive, trying to reorganise the remnants of the student network but failing, and in desperation making contact with the Communist guerrillas whose base camps were in remote mountains. 'But always I was thinking,' he said quietly, 'always thinking of you, and of being with you, Jinda.'

The moon had risen without either of them noticing, and the pale moonlight filtered down from the canopy of willow leaves. Ned stopped talking, and leaned over to touch Jinda's cheek with his fingertips. Jinda sat very still, as Ned stroked her face, her neck, then her bare arms.

'Touch me,' Ned whispered, so softly that it sounded like the rustle of leaves overhead. 'Touch me, Jinda,' he said again.

He took off his shirt, and his bare chest gleamed in the moonlight. Cautiously, Jinda reached out and touched him.

His chest was cool and hard. Like the slabs of stone on the river bed, polished smooth by years of running water, she thought. With both hands, she touched him, her fingers fanning out along the delicate ridge of his collar bones, and over his chest.

'So hard,' Jinda whispered. How lean and strong a man's body was, she thought. Hard where a woman was soft, like the plough against the soft moist earth. 'When the rains come, you can plough our land,' Jinda said. She let Ned hold her wrists and lift her palms towards his mouth. He kissed each hand, and she felt a strange, tingling sensation. 'And a spice garden,' she said dreamily. 'And a new thatched roof, and children too.'

Ned held her hands away from him a little, and looked at her. 'What're you saying, Jinda?' he said.

'The life we're going to share,' Jinda said. 'Why're you

looking at me like that? Don't you want children? Fat little babies with . . .'

Ned let go of her hands, and took a deep breath. For a moment he said nothing. Then he shook his head and said, 'Our life together isn't going to be full of fat babies and spice gardens, Jinda,' he said carefully. 'At least mine isn't.'

'What do you mean?'

'Haven't you guessed? I'm on my way to join the Thai Communists. Thousands of other students have joined, and . . .'

'What about me?' Jinda asked, her voice choked.

'I hope that you'll come with me. That's why I came, Jinda, to ask you to join me. You said you wanted to build a life with me . . .'

'Yes, but . . . but, I meant together — at Home!'

'In Maekung?'

'Yes, in Maekung,' Jinda said, sitting up straight now and talking very earnestly. 'We can plough the fields together, and plant this year's rice. The rains will come, I know they will. And after the harvest, we could . . .'

'We could hand over half of our crop to the landlord?'

'We'll still have the other half,' Jinda said, imploringly. 'Maybe someday it'll change, but . . .'

'Someday! Someday will never come unless we fight for it, Jinda. Hasn't the last few months taught you anything? We've got to fight for what we want.'

Jinda's heart sank. 'Fight? You mean with guns?'

'It has come to that,' Ned said.

'You'd shoot a gun, wound people, kill them?' Jinda's voice shook. 'Like they did to us, at the rally that day?'

'Believe me, Jinda, I don't want to,' Ned said. 'But I've tried peaceful means. Liberal dissent, parliamentary reform — all the things that're supposed to work in a democracy — where did it get us? The military took over without our even putting up a struggle. Just like that, they're back in control of the government. Don't you see, Jinda? The only thing the military understands is force. They shot at us; we've got to shoot back at them. It's that simple.'

Jinda shook her head. 'Killing people,' she said miserably, 'is never "that simple".'

'You'd never have to shoulder a gun, Jinda,' Ned said quickly. 'Especially since Sri's already taught you some basic first aid. You could help nurse the wounded. There're always wounded guerrillas after every skirmish with the government soldiers . . .'

'I've had enough of fighting!'

'But don't you see? There's no other way,' Ned said. 'Violence is the only way to overthrow the ruling class and achieve justice and equality . . .'

'Justice!' Jinda said fiercely. 'Can you taste justice? Can you smell equality? What do all your fancy words mean? I can't live my life for things that I can't taste, or smell, or hold in my hands. Soil after the rain, it has such a rich, sweet smell. And tamarind shoots leave a golden taste on your tongue. These things are real. These things I can live for.'

'But Jinda . . .'

'No, let me finish. I don't want to see any more bloodshed. My father died before his time, for these empty words. I don't want . . .'

'Your father died,' Ned broke in angrily, 'fighting for a new rent law. I've got to fight on so that he didn't die in vain. Can't you understand that? Doesn't his death mean anything to you?'

'Yes, it does!' Jinda replied, trying to keep her voice steady. 'It means I don't want to see you dead too — and all for some vague dream! It means I want you to live with me, to work the soil that my father worked before, and yes, to raise some fat babies. I want to live, and grow things, and be happy. Is that so wrong?'

Jinda was not prepared for the tears that rolled down Ned's cheeks, leaving shiny streaks that reflected the moonlight. 'No, that's not wrong,' he said, very softly. 'I want that too, very much,' and his voice was so filled with longing that Jinda felt her eyes sting too. 'But I'm not ready for that yet, Jinda. There is too much I must do first.' He turned away, hugging his knees to him.

He looked thin and fragile, and yet, Jinda realised with a slow wonder, he had a kind of strength nothing could snap.

Slowly, Jinda lay down on the moss. Through the lacy mesh of willow leaves above her, a crescent moon shone. Only the whir of the cicadas broke the utter stillness of the night.

Softly, Jinda called to Ned. 'We're both tired,' she said. 'Come sleep now.' She watched as he uncurled his long arms and legs, and lay down carefully next to her. He stretched out an arm, and drew her to him. She put her head on his shoulder, and snuggled against him. Through her thin blouse she could feel his warmth, and she pressed herself against that. How comfortable, she thought sleepily, those puppies on his porch must've been!

Jinda awoke with a start. Had she imagined it, or was that thunder rumbling in the distance? A bolt of lightning flashed in the west, followed by another roll of thunder. Jinda's pulse quickened. Suppose the monsoon rains would start tomorrow — how would they manage to do everything in time? There was the ploughing to be finished, then the harrowing and mending of dykes, and the transplanting. If only Ned would come back with her and help.

Jinda looked down at Ned wistfully. He was still asleep, his arms flung up over his head, his chest rising and falling evenly. He looked young and very vulnerable. Gently she touched his cheek. He stirred, then, blinking, looked up at her.

'It's morning?' he asked drowsily.

Jinda nodded.

'So soon,' he said.

'So soon,' Jinda repeated sadly.

They did not say anything again for some time. The dawn light sifted through the leaves, dappling a starling perched on a branch above them. When Ned stood up, it flew away, soundlessly.

'Won't . . . won't you change your mind?' Jinda asked.

Ned smiled. 'Won't you?' he answered. He picked up the knapsack he had with him, and slung it over his shoulder.

'If . . . if you ever want to come back,' Jinda said, 'you know I'll be here.'

He looked at her for a long time. Then he said, gravely, 'I know that, Jinda.' Without looking at her, he turned and walked away.

She watched him, climbing up a steep craggy path on the mountainside. Lithe and sure-footed, Jinda thought, like a leopard. She felt a surge of pride in him. Maybe he's wrong, Jinda thought, maybe he's headstrong and foolish to fight for words that most of us don't even understand. But he's young and brave, and maybe he'll actually win.

He reached the top of the hill, and turned around. Silhouetted against the morning sky, he lifted one arm in an expansive farewell. Jinda tried to smile, but it was a fluttery, weak smile. In any case, she knew he was already too faraway to see it.

Chapter 15

After Ned disappeared over the crest of that hill, Jinda's first impulse was to chase after him, and go with him after all. She started to run, then stopped abruptly. What for? He didn't need her. He'd welcome her, of course, but he didn't really need her. Not the way Dao did, and little Pinit and her grandmother did. Not the way the land, that dry, barren land did.

Dark storm clouds were gathering on the horizon, and a brisk wind seemed to be blowing them towards the valley.

If only the rains would start, Jinda thought.

Slowly she walked back towards the village. Near her house, she came upon a group of children, playing clay marbles in the sand. A marble rolled near her toe, and as she bent to pick it up she heard a scream.

'That's Dao!' Jinda cried. She turned and ran towards her own house. When she had climbed up the ladder and reached the doorway of the inner room, she stopped.

Dao had crawled off the sleeping mat and, on her hands and knees, was keening loudly, her whole body rigid. Her hair had fallen over her face, and her clothes were in disarray. She swayed to and fro, her back arched. Then her moans subsided, and she sank back onto the floor.

Jinda tiptoed into the room and approached her sister. 'Dao,' she murmured.

Her sister's eyelids fluttered open. 'You're back,' she whispered, managing a weak smile. 'And Ned?'

Jinda smoothed back some strands of Dao's hair. 'He's gone,' she said. 'He wants to fight, and I,' she shrugged, 'I want to grow things.' She could feel a sharp sting in her eyes, and hastily she took a damp cloth and gently wiped Dao's face with it.

Their grandmother, sitting by the window, watched them. 'I always said that boy had strange ideas,' she told Jinda, but her eyes were gentle. Then Dao moaned again, and the old woman quickly became more businesslike.

'Cut me a stack of fresh banana leaves,' she said to Jinda briskly. 'And get me that big kettle of boiled water. One more thing, cut me a sliver of bamboo, about the length of a kitchen knife blade. But be sure to split it carefully, without letting anything touch its edge as you split it. Understand?' She edged over to help support Dao as she started moaning again.

Glad to have something to do, Jinda left the room. Outside, the wind had increased, and the broad flat banana leaves flapped wildly about the spongy trunks as Jinda slashed off leaf after leaf. When she had about six leaves tucked under her arm, she brought them back to the house verandah.

Then she set off to the bamboo grove, and chose a clean, brittle pole growing to the side of the grove. It was tough, but after a brief struggle Jinda sliced through it, then ripped a vertical segment off. Cleaner than any metal knife, untouched by human hand, this sliver of bamboo would be used to cut the baby's umbilical cord.

The wind was so fierce now that it whipped the sharp bamboo leaves in her face, stinging her cheeks. Head bent,

Jinda pushed her way out of the grove and stumbled back to the house. Heavy clouds of dust and sand swirled about her ankles. From neighbouring houses came the sound of wooden shutters being slammed shut against the wind and sand.

Would the wind blow the rainclouds right past them, Jinda wondered? Squinting against the dust, Jinda scanned the sky. But another shrill cry came from the house, and Jinda hurried back.

Dao was in the middle of another contraction. Gripping the pillows on either side of the mat, her arms anchored her writhing body down to the floor. The veins bulged in her neck, as she moaned. Then the spasm of pain passed, and Dao went limp.

Through the rest of the morning, Jinda cradled her sister's head in her lap, brushing back the hair from her eyes, feeding her sips of cool well-water, and massaging her back and legs when she wasn't in pain.

Once a gust of wind blew a clump of leaves through the window, flinging them in a swirling eddy against the walls. Dao grabbed one, and crushed it into tiny fragments.

In the late afternoon, as the last painful contraction had passed, her grandmother knelt down and massaged Dao's legs. 'It's almost over, child,' she said. 'Rest when you can, there, close your eyes.' Soothingly, she talked until Dao's breathing became more regular.

The pain began again. Jinda saw Dao tense herself for another contraction, but it never came. Instead, Dao struggled to sit up. The old woman pushed her back down gently.

Dao grunted, straining. 'I've got to . . .'

'Push, right? You want to push it out — go ahead. Push, woman!' her grandmother said.

Grimacing, Dao grabbed onto her sister's hands and pushed. Her breathing came in sharp, erratic pants, and she squeezed her eyes shut as she strained.

Deftly, the old woman undid the knot of Dao's sarong, and draped the cloth up over Dao's knees. Then she ducked down, her face disappearing between Dao's spread legs. When she reemerged over the edge of the sarong, she was smiling.

'It's coming,' she said. 'I can see a bit of its head. You're almost there, Dao. Hold on tight, there, and push again!' The old woman calmly spread out the stack of powdery smooth banana leaves under Dao's buttocks and legs, so that there was a double lining of leaves over the mat.

Dao hung onto to Jinda's hands. Low long grunts seemed to well up from deep within her as she squeezed the baby downwards.

'I can't . . . oh, I can't . . .' she gasped.

Abruptly she let go of Jinda's hands, and pressed down hard against her own belly, as if trying to block the surge outwards. 'Stop it,' she cried, '. . . too big!'

But her grandmother pried Dao's hands away, and placed her own hands on the heaving womb. Instinctively, she sensed when every new muscle contraction would crest, and she squeezed Dao's abdomen, slowly and gently pressing on the mound.

'Jinda,' she murmured. 'Spread her legs wider. Go on, child, this is no time to be shy! And lift her sarong up more. Can you see the baby's head?'

Hesitantly Jinda looked. Between the vertical folds of her sister's opening, she saw what seemed to be a patch of slick black hair. Each time Dao pushed, the folds of flesh were pulled wider apart, like reluctant curtains being drawn, and a fraction more of the slick patch was revealed.

'I can see it,' Jinda said breathlessly.

It looked huge, rammed against Dao's torn opening, and still it continued to expand. Rounder, fuller, it pressed outwards. And then suddenly it was out, domed head completely through, its slender neck sprouting between Dao's legs like a stalk. Before Jinda could reach for it, the rest slithered out — chest, arms and legs. Slim and slippery and blue-grey, it looked like a catfish squirming out of a mud puddle.

Jinda touched it. It was warm and wet. Wet hair plastered against its forehead, eyes shut, its tiny fists clenched. A fighter, Jinda noted with some satisfaction, Awkwardly she held it up towards Dao, her hand cupping its smooth moist buttocks.

'A girl,' Jinda heard her grandmother say to Dao.

She looked down at the creature in her hands. Yes, it was already a tiny replica of its mother, complete with a tiny set of petal-like folds like the ones from which she had just emerged. Jinda was glad it was a girl.

Her grandmother was kneading Dao's stomach now, with a firm, vigorous rhythm. Dao groaned, and flung out her arm to brush the old woman's hands off, but it was a feeble gesture.

As the grandmother continued to press, the afterbirth was delivered. A round red slab like calf liver, it looked engorged and slick. The air reeked of fresh blood, and Jinda felt sick. With the placenta, a gush of blood poured down Dao's thighs, forming a dark red puddle on the banana leaves. Dao's feet and sarong were streaked with blood, yet with every push of the old woman's hands, another spurt of blood would flow out.

For a split second images of the student massacre flashed across Jinda's mind. Blood-smeared thighs, bodies being dragged across the red-streaked grass, and that shoeshine boy. Jinda stared at her bloodstained hands, immobilized.

'Help me,' her grandmother was saying. The old woman crawled towards her. Fumbling, she took out a piece of thread and tied a knot around the pulsating umbilical cord, inches away from the baby. Her gnarled hands were trembling badly.

'You cut,' she said, handing the bamboo sliver to Jinda.

Jinda felt the piece of sharp bamboo being thrust into her hand.

'Right here,' the old woman said, pointing to the knot she had just tied. 'On this side of the knot. Cut!'

Jinda took a deep breath, but the sight of blood made her dizzy. She shook her head at her grandmother. 'I . . . I can't . . .' she said hoarsely.

And then the baby cried. It was a triumphant sound, strong and sweet with the pulse of life. There is the blood of death, and the blood of life, Jinda suddenly realised. And this blood on my hands is the blood of life.

She looked down at her own hand, and saw that it held the bamboo blade without a tremor. Quickly, she bent down and slashed through the cord, still turgid with the flow of the mother's blood to the child. The cut was clean and sharp.

The baby wailed.

'Listen to her!' Jinda called to her sister.

Dao was listening, but not to the baby. Startled, her eyes wide with wonder and shock, she was staring out of the open window.

Only then did Jinda hear it too; a quick frantic hammering all around them. The roof, the walls, the leaves outside were trembling with the sound. The whole room pulsed with it. Unbelieving, holding her breath, Jinda looked out of the window.

It was raining.

'She's brought the rain,' Dao said, holding out her arms for the baby now.

Jinda held the slippery little creature towards her. She whimpered, waving one tiny fist in the air.

'Wash her,' Dao said softly.

'We will, child,' the grandmother said. 'You lie down. Rest.'

'Wash her,' Dao insisted. 'Now. In the rain.'

And so Jinda took the baby in her arms, a damp limp thing, and walked out. Fresh gusts of wind blew under the eaves, tearing bits of thatching with it.

Jinda stepped out from the shelter of the eaves and onto the open verandah, holding the baby close to her.

Daughter of the rain, she said silently, here — meet the sky.

The rain was warm and gentle, coursing down Jinda's face and onto the tiny baby. She watched the raindrops stream over its smooth skin, rinsing away the smear of blood.

At the top of the steps she sat down, holding the baby in her lap, where a pool of rainwater slowly collected in her sarong. With one hand she held the baby's limp neck, and with the other she scooped the rainwater from her lap to bathe the baby.

Other villagers were tumbling out of their houses, laughing up at the sky. Children tore off mud-caked clothes and ran naked in the warm rain, their bare buttocks slick and brown. Housewives set out limp herb gardens in woven baskets under the sky, while men pushed glazed urns under bamboo drains from the roofs to catch the rain. And in a corner, an old man stood alone, his face uplfted to catch the raindrops on his tongue.

The water in her lap became tinted a pale red from bloodstains, but Jinda did not mind. Then the rain started coming down harder, stinging her scalp and cheeks. The baby started to cry.

'Give her to me!' Jinda's grandmother called from the doorway.

'But she's enjoying the rain!' Jinda called back.

'She'll catch a cold! Give her to me,' the old woman said. 'Now!'

Laughing, Jinda handed the baby, now glistening clean, back to her grandmother.

'Is Dao all right?' Jinda asked, as together they wrapped

the baby tightly into a thick blanket the old woman had brought with her.

'She's resting,' her grandmother said.

'So . . . you don't need any more help?'

The old lady smiled, shaking her head. 'Go ahead,' she said. 'Go and run in the rain. If it weren't for the baby, I'd even join you.'

Jinda slithered down the steps of the verandah and ran towards the swinging bamboo gate. The earth was soft and pliant under her bare feet. How good to feel mud again, Jinda thought.

Out in the lane, the wind was stronger, and a gust slashed against her arms like long sharp rice stalks. She passed the hibiscus hedge, and saw its leaves streaked a bright green where the rain had washed off the thick layers of dust. A few flowers burst through, cleansed a brilliant scarlet by the rain.

Sprinting down the lane, Jinda's steps were light and springy against the mud-slick soil. She felt the wind against her face, whipping her hair behind her, piercing through her thin wet clothes so that they felt like a second skin on her.

The storm was building up momentum. Wind that had been gentle just moments ago now blew so strongly that the trees in its path strained at their roots. Lashing out wildly, the palm fronds swerved and swooped wildly against the sky.

And still the rain quickened, pelting at the flat earth from the cloud-flattened sky.

Through the storm Jinda ran on. Running, she gulped in great mouthfuls of the wind, and felt the rain soak her clothes to her skin. Running, she felt the hard dry knot within her uncoil and grow soft and pliant again. She ran until her sides hurt, and her breath came in painful gasps.

The village lay far behind. Before her the rice fields

stretched out, reaching to the foot of the mountains. Bathed in wet shadow, the mountains glowed a vibrant lilac.

A few farmers were already out in the fields, yoking their buffaloes to start the ploughing, and a few women were wading knee-deep in their seedbeds, straightening the seedlings that had lodged in the rain.

She saw Pinit, splashing in the fields, rattan fish-basket strapped across his shoulder. Together with a group of other naked boys, he stalked the little ditches now overflowing with rainwater. There might be tiny translucent shrimp for dinner tonight, and a delicate frog or two.

Pinit saw her, and waved, his rib cage etched sharply under his skin. How thin he has grown, Jinda thought with a stab of pity. She waved back to him, but did not feel ready to join him, or any of the exuberantly busy villagers out in the fields.

Slowly she walked at the edge of the fields, until she found herself near the cremation grounds. In the clearing of that grove of Bodhi trees, where her father had been burned to ashes, Jinda hesitated, then walked into the grove.

It was deserted and dark, yet Jinda felt strangely drawn to the place. The thick canopy of leaves sheltered her from the rain. Shadows quivered under the leaves, and she walked gingerly at first, as if afraid to intrude. She felt as if she had stepped into some intensely private sphere, where the silence was a shield enclosing the space within from the world outside.

She stood next to the same Bodhi tree she had leaned against during the funeral, as she waited for her father's coffin to catch fire. Jinda shut her eyes. She thought of her father, tried to remember him not as that still, gaunt figure lying in the coffin, not as flakes of ash wafting through that hot afternoon, but as the man before that. He had taught her how to weave fish traps out of rattan strips, and how to make a kite, and how to hold a sickle. He had been tall, his eyes crinkled at the corners when he smiled, and he had smelt of woodsmoke.

But the image would not form. Jinda saw only his hand as it had cupped the clear coconut water in the coffin, the palm horribly swollen and scarred. Not this hand, Jinda thought desperately, that was not the hand that touched me and held me. His hand was strong.

And then she remembered.

His hand is big and bony, with hard bumps on his palm. His fingernails are stained a reddish brown from working in the soil, and his big thumb is just the right size for holding onto. It is this thumb she clings onto as they cross the fields. Balancing carefully on the narrow dykes, she holds onto his hand so that she does not fall.

He walks fast, but steadily. He never falls. The brown rice stalks bow on either side of her as a breeze sifts through. Their heads of grain droop above her head. Rows and rows of brown stalks, stretching out endlessly.

'Why are they all brown?' she asks.

'What?' his voice is dreamy, faraway, way above her head.

'The stalks. Why do they turn so brown? Why don't they stay that pretty green?'

Nothing stays pretty forever,' he says quietly. 'When they're young and strong, they're green. But when the rice grains form, the stalks send all their strength up to the grain. So as the seed grow, the stalks dry out, and turn brown.'

'Poor stalks!'

He laughs, and his laughter stirs her like a cool breeze, making the back of her neck tingle. 'Don't feel sorry for the stalks, little one. They don't mind. The seeds they've fed will grow into rice stalks someday, and those stalks will die to make more new seeds.'

'Why?'

'That's how life goes on. The old must give up their strength so the new can grow.'

'Why?'

'That's just the way things are.'

'But why?'

'Why, why, why!' He laughs and lifts her up, tosses her up in the air. For one heartstopping moment she is sailing above the brown fields, flying towards her father's strong, waiting arms. 'Why? Because I love you, that's why!'

Because I love you, she echoes, and laughs too.

The tears streamed down Jinda's cheeks now, merging with the rain. She wept hard, but she was laughing too, and it was a relief to laugh and cry at the same time. Hugging the Bodhi tree, her body shook so hard she could almost feel the great tree tremble slightly, deep down in its roots.

Of course he hadn't died for an idea. Ned was wrong, her father hadn't died for justice, or equality or democracy. He had given up his life for little Pinit, and Dao, and Her, so that they might grow stronger. And he had done it out of love.

Jinda wiped her cheeks, and walked back into the grove. At the centre of the ring of trees was the clearing, where the twin walls of brick had supported the coffin during the cremation. The mortar had crumbled in places, and a few of the bricks had fallen loose. Otherwise it was intact.

Jinda walked up to the brick wall, and rested her hands on it. There was a thin layer of black ash on the rough brick. Slowly she ran her finger across it, wiping off a thin film. Green plants to brown stalks, she thought, and a strong man to black ash.

She squatted down between the walls, and scooped up a heap of ash that had accumulated there. The rain had dampened the outside layer, but deep inside the pile, the ash was still dry and loose. Carefully Jinda brought out a handful. Each flake was fragile, paper-thin.

Cupping the ash in both hands, Jinda carefully walked out into the fields. The storm was clearing, and thin streaks of light glimmered between the rainclouds. Two swallows darted out from some nearby trees, shaking tiny sprays of rain from their forked tail-feathers. In the distance the mountains shimmered in a haze of blue.

For a moment Jinda stared at the mountains. A curtain of rain hung suspended over some of the higher peaks. Ned might be there now, Jinda thought, feeling the rain on his skin too. Perhaps he might feel the call of the fields in the afterrain, someday, and lay down his weapons to come and look for her? The mountains stretched out, range beyond range, far into the distance.

A few villagers called out to her, and she smiled. But she avoided them, skirting the fields where people were ploughing, until she had reached her seedbed.

There it was, the patch of bright green seedlings, quivering in the rain.

Jinda stepped down from the bund and into the muddy water of the seedbed. The water swirled cool and clear around her ankles. What a wonderful rain it was, she thought! Even those seedlings which were shrivelled were reviving now, their stems upright and taut. The sea of green stalks rippled gracefully in the wind, beautiful and alive.

'Father,' Jinda said aloud, and scattered her handful of ashes on each of the seedlings. Tiny flakes of ash drifted through the rain-fresh air, and landed on the surface of the water, gathering in little clusters at the base of the seedlings.

Delicate but resilient, the tiny stalks glowed a fresh young green. If the rains continued in the next few days, the fields could all be ploughed and harrowed, and the seedlings finally transplanted. Choked for space now, they would thrive once transplanted in the moist, muddy fields.

Jinda looked at them. Already the flakes of ash were sinking into the water. Soon they would merge into the moist earth, and be absorbed by the roots of these seedlings.

All as it should be, Jinda thought, and smiled. Maybe we will have a good harvest next year, after all.

General Editors: Anne and Ian Serraillier

Chinua Achebe Things Fall Apart
Vivien Alcock The Cuckoo Sister; The Monster Garden; The Trial of Anna Cotman
Michael Anthony Green Days by the River
Bernard Ashley High Pavement Blues; Running Scared
J G Ballard Empire of the Sun
Stan Barstow Joby
Nina Bawden On the Run; The Witch's Daughter; A Handful of Thieves; Carrie's War; The Robbers; Devil by the Sea; Kept in the Dark; The Finding; Keeping Henry
Judy Blume It's Not the End of the World; Tiger Eyes
E R Braithwaite To Sir, With Love
F Hodgson Burnett The Secret Garden
Ray Bradbury The Golden Apples of the Sun
Betsy Byars The Midnight Fox
Victor Canning The Runaways; Flight of the Grey Goose
John Christopher The Guardians; Empty World
Gary Crew The Inner Circle
Jane Leslie Conly Racso and the Rats of NIMH
Roald Dahl Danny, The Champion of the World; The Wonderful Story of Henry Sugar; George's Marvellous Medicine; The BFG; The Witches; Boy; Going Solo; Charlie and the Chocolate Factory
Andrew Davies Conrad's War
Anita Desai The Village by the Sea
Peter Dickinson The Gift; Annerton Pit; Healer
Berlie Doherty Granny was a Buffer Girl
Gerald Durrell My Family and Other Animals
J M Falkner Moonfleet
Anne Fine The Granny Project
F Scott Fitzgerald The Great Gatsby
Anne Frank The Diary of Anne Frank
Leon Garfield Six Apprentices
Graham Greene The Third Man and The Fallen Idol; The Power and the Glory; Brighton Rock

Marilyn Halvorson Cowboys Don't Cry
Thomas Hardy The Withered Arm and Other Wessex Tales
Rosemary Harris Zed
L P Hartley The Go-Between
Esther Hautzig The Endless Steppe
Ernest Hemingway The Old Man and the Sea; A Farewell to Arms
Nat Hentoff Does this School have Capital Punishment?
Nigel Hinton Getting Free; Buddy; Buddy's Song
Minfong Ho Rice Without Rain
Janni Howker Badger on the Barge; Isaac Campion
Monica Hughes Ring-Rise, Ring-Set
Shirley Hughes Here Comes Charlie Moon
Kristin Hunter Soul Brothers and Sister Lou
Barbara Ireson (Editor) In a Class of Their Own
Jennifer Johnston Shadows on Our Skin
Toeckey Jones Go Well, Stay Well
James Joyce A Portrait of the Artist as a Young Man
Geraldine Kaye Comfort Herself; A Breath of Fresh Air
Clive King Me and My Million
Dick King-Smith The Sheep-Pig
Daniel Keyes Flowers for Algernon
Elizabeth Laird Red Sky In the Morning
D H Lawrence The Fox and The Virgin and the Gypsy; Selected Tales
Harper Lee To Kill a Mockingbird
Laurie Lee As I Walked Out One Midsummer Morning
Julius Lester Basketball Game
Ursula Le Guin A Wizard of Earthsea
C Day Lewis The Otterbury Incident
David Line Run for Your Life; Screaming High
Joan Lingard Across the Barricades; Into Exile; The Clearance; The File on Fraulein Berg
Penelope Lively The Ghost of Thomas Kempe
Jack London The Call of the Wild; White Fang
Lois Lowry The Road Ahead; The Woods at the End of Autumn Street
Bernard Mac Laverty Cal; The Best of Bernard Mac Laverty
Margaret Mahy The Haunting; The Catalogue of The Universe
Jan Mark Thunder and Lightning; Under the Autumn Garden

How many have you read?